Manipulation of Groundwater Colloids for Environmental Restoration

Manipulation of Groundwater Colloids for Environmental Restoration

Edited by
John F. McCarthy
Frank J. Wobber

Associate Editors
**Edward J. Bouwer, H.Scott Fogler, Philip M. Gschwend,
Ronald W. Harvey, James R. Hunt, Richard G. Luthy,
John C. Westall, and John M. Zachara**

Editorial Assistant
Cindy T. Woodard

Proceedings of a Workshop in Manteo, North Carolina
October 15-18, 1990

Sponsored by the Subsurface Science Program,
Environmental Sciences Division
U.S. Department of Energy

LEWIS PUBLISHERS
Boca Raton   Ann Arbor   London   Tokyo

Library of Congress Cataloging-in-Publication Data

Catalog record is available from the Library of Congress.

ISBN 0-87371-828-3

PRINTED IN THE UNITED STATES OF AMERICA
1 2 3 4 5 6 7 8 9 0

Printed on acid-free paper

John F. McCarthy is a Senior Scientist in the Environmental Sciences Division of the Oak Ridge National Laboratory (ORNL). His current research focuses on the transport of colloids and natural organic matter in subsurface systems, and their effect on the mobility of contaminants in groundwater. He is coordinator of a subprogram on colloid research within the Subsurface Science Program of the U.S. Department of Energy (DOE), and has organized a series of international symposia on this subject for DOE. Dr. McCarthy received a B.S. from Fordham University in 1969 and a Ph.D. in oceanography from the University of Rhode Island in 1974. From 1975-79 he was a postdoctoral fellow and research associate at the Biology Division of ORNL, and joined the Environmental Sciences Division in 1980. His research interests include contaminant fate and transport, as well as the biological availability of contaminants, with an emphasis on the use of molecular and biochemical markers of exposure and effects in environmental species. He has published over 75 papers and edited a book in areas of environmental toxicology and chemistry. He is a member of scientific advisory committee for the United Nations Environmental Program's Marine Mammal Action Program.

Frank Wobber is currently Program Manager of the U.S. Department of Energy's (DOE) Subsurface Science Program. He is responsible for long term, basic research in a variety of areas including the chemistry and microbiology of organic-radionuclide mixtures, colloids and biocolloids, and immiscible fluid flow. Dr. Wobber initiated the Department's Deep Microbiology Program which is focussed on exploring the microbial ecology of deep sediments and aquifers, and in the origins of the deep microbiota. These research programs emphasize interdisciplinary research in hydrogeology, microbiology and geochemistry. He is active in various interagency committees dealing with ground water, including the Committee on Earth and Environmental Science Subcommittee on Water Resources, and in departmental activities in environmental restoration at DOE sites. Dr. Wobber received a M.S. Degree in Geology from the University of Illinois, and a Ph.D. from the University of Wales as twice Fulbright Scholar to the United Kingdom.

PREFACE

An interactive research conference series on colloids in the environment was initiated in 1984 as part of the Department of Energy's (DOE's) Subsurface Science Program. Since 1984, three meetings have been organized jointly by DOE and Oak Ridge National Laboratory (ORNL). DOE/ORNL conferences on colloids in the environment are held in the town of Manteo on Roanoke Island in eastern North Carolina. The meeting site was originally selected because of its proximity to a field research site at Engelhard, North Carolina, where exploratory research on the origin and transport of organic colloids was in progress. Since 1984, these biannual scientific meetings (1986, 1988, and 1990) on colloids have become identified with their location and are referred to as "Manteo I, II, and III." The first two conferences focused on the scientific research needs associated with colloid formation and migration; fundamental questions about colloid-associated contaminant transport were also explored. These two meetings contributed to (1) refining the future directions of DOE's basic research program in colloids and (2) preparing a 5-year research plan that was published in 1988. These biannual conferences have evolved into an international forum for scientists and engineers with research interests in the mobility and stability of inorganic and organic colloids (including biocolloids) in the natural environment and, particularly, in subsoils and groundwater. International participation, an emphasis on critique of scientific ideas provoked by formal presentations, and the preparation of a summary volume, characterize these conferences.

This report summarizes the scope and the results of Manteo III, *Concepts in Manipulation of Groundwater Colloids for Environmental Restoration.* The conference that was held in Manteo, North Carolina, October 15-18, 1990. Representatives from governmental agencies and universities in Australia, France, the Netherlands, Great Britain, and Canada participated; 40 to 50 representatives from scientific groups in the Unites States also attended. The diverse scientific skills represented by these attendees ranged from fundamental colloid chemistry to the hydraulics of colloids at waste disposal sites. The gaps that commonly exist between basic and applied research areas and between laboratory and field studies were also bridged.

The goal of this conference differed from previous meetings in that opportunities to take advantage of organic and mineral colloids, biocolloids, and colloidal complexes to facilitate cleanup of contaminated subsurface systems were explored. Previous conferences focussed on issues for which some research experience existed; scientific papers on the theory of colloid chemistry, colloid migration, or colloid-associated contaminant transport are available. However, "the book has yet to be written" on the extent to which colloids can be used in the long term for environmental restoration. Manteo III participants analyzed colloids in natural subsurface systems as a positive force for environmental restoration rather than as problems to be resolved. DOE did not anticipate that new engineered technologies would emerge from the meeting or that engineering feasibility would be assessed; rather, the objective was to identify concepts that in the long term might complement existing engineered cleanup technologies by modifying natural subsurface conditions (e.g., manipulation of aquifer redox chemistry) in situ. In this sense, these proceedings represent a compendium of new ideas, and a new source of information on the potential use of colloids as contributors to environmental restoration.

DOE's Subsurface Science Program conducts basic, long-term research relevant to DOE's long-term concerns about environmental restoration and reducing cleanup costs at DOE sites. The conference provided DOE with the basis for evaluating potential scientific opportunities for using organic and inorganic colloids for site restoration. Additionally, DOE gained a perspective on concepts for cleanup that will complement rather than substitute for established engineering methods.

DOE and ORNL acknowledge the scientific contributions of all attendees and especially thank session moderators and discussion leaders who contributed to Manteo III's success.

Frank J. Wobber, Program Manager
Subsurface Science Program
Environmental Sciences Division
U.S. Department of Energy

and

John F. McCarthy
Environmental Sciences Division
Oak Ridge National Laboratory

ACKNOWLEDGMENTS

The active participation of the speakers and all the attendees is gratefully acknowledged. Special thanks are due Jeannette Cox for helping to arrange many of the logistical details of the meeting and for acting as our hostess during the meeting. Thanks also to Mr. Rhett White and the staff of the North Carolina Aquarium in Manteo, North Carolina, for providing the meeting rooms and assisting with many of the logistical details of the meeting. Support for some meeting costs were provided by the Environmental Science Research Center, Pacific Northwest Laboratory.

The meeting was sponsored by the Subsurface Science Program, Environmental Sciences Division, Office of Health and Environmental Research, U.S. Department of Energy. The Oak Ridge National Laboratory is managed by Martin Marietta Energy Systems, Inc., under Contract No. DE-AC05-84OR21400 with the U.S. Department of Energy.

ABSTRACT

A meeting on "Conceptual Strategies for Manipulating Groundwater Colloids for Environmental Restoration" was held on October 15-18, 1990, in Manteo, North Carolina. The meeting was part of a series of research conferences on colloids in subsurface environments initiated in 1984 as part of the Subsurface Science Program of the U.S. Department of Energy's Environmental Sciences Division, Office of Health and Environmental Research.

The purpose of the meeting was to evaluate the question: Can a fundamental understanding of processes controlling the mobilization and deposition of groundwater colloids be applied to the development of strategies to manipulate colloid mobility for mitigation or remediation at waste sites? The objective was to examine the current state of science with respect to the feasibility of such approaches and to identify and prioritize fundamental research needs (both in the laboratory and the field) that will be required to develop these capabilities within the next 5 to 20 years. The scope of the meeting included consideration of both organic and inorganic colloids and organic, heavy metal, and radionuclide contaminants, as well as mixed wastes.

Technical sessions included:
1. Mobilizing and Depositing Colloids (Can chemical modification of subsurface systems be used to reliably manipulate colloid stability in situ?)
2. Biocolloid Mobility (What the mobility of microorganisms in porous and fractured media be manipulated to improve delivery of microbes for bioremediation?)
3. Manipulating Colloids to Change Aquifer Permeability (Will changes in colloid stability affect the physical structure of a formation to either enhance bioremediation or create in situ barriers?)
4. Introducing Modified Colloids, Surfactants and Emulsions (Can surfactants or modified colloids be used to control colloid mobility or to increase recovery of sorbed contaminants by pump and treat methods?)

Discussions at the meeting are a measure of the disciplinary diversity of participants. Topics of discussion ranged from research on the mobility of organic macromolecules by controlled field injection experiments to new techniques to investigate the surface chemistry and aggregation of inorganic colloids. Other topics included the depositional behavior and transport of biocolloids (microorganisms) in porous media, surfactants as modifiers of surface binding sites on colloids, and genetic engineering of microorganisms to serve as a contaminant-scavenging biocolloids. Although the exploitation of colloidal particles for remediation has yet to be fully defined, several colloid remediation scenarios were identified during the conference:

- Contaminant scavenging, in support of pump and treat methods.
- Pore clogging to reduce permeability as a means to limit the mobility/ release of contaminants from a source term or to isolate subsurface formations in advance of treatment.
- Introduction of nonindigenous colloids, specialized biocolloids, or surfactants to optimize contaminant scavenging, reduce permeability of a formation, or enhance bioremediation.

This document summarizes the scientific results of the meeting with the goal of identifying possible new and innovative manipulation strategies and concepts, evaluating the strengths and limitations of each concept, and identifying the long-term research that might be required to refine these concepts. Technical sessions included (1) analyses of subsurface chemical and microbiological conditions that, if modified, could facilitate colloid mobilization and deposition in situ; (2) controls on the mobility of biocolloids (microorganisms) in porous and fractured media; (3) manipulation of colloids to modify aquifer permeability (i.e., to increase permeability or to clog pores); (4) surfactants to control colloid mobility or increase recovery of colloid-associated (sorbed) contaminants by pump and treat methods (i.e., as complements to engineered technology); and (5) the influence of interacting chemical, microbial, and hydrologic processes on proposed manipulation strategies.

LIST OF MODERATORS,
DISCUSSION LEADERS AND SPEAKERS

Dr. Roger Bales
Department Hydrology and
  Water Resources
University of Arizona
Tucson, AZ 85721

Dr. Steven A. Boyd
Michigan State University
Department of Crop and Soil
  Science
East Lansing, MI 48824-1325

Dr. Ed Bouwer
Department of Geography and
  Environmental Engineering
The Johns Hopkins University
34th and Charles, Ames Hall
Baltimore, MD 21218

Dr. Al B. Cunningham
Center for Interfacial Microbial
  Process Engineering
Montana State University
409 Cobleigh Hall
Bozeman, MT 59717

Dr. H. Scott Fogler
Department of Chemical
  Engineering
374 Dow Building
University of Michigan
Ann Arbor, MI 48109

Dr. D.W. Fuerstenau
Department of Materials Science
  and Mineral Engineering
University of California
Berkeley, CA 94720

Dr. Philip M. Gschwend
Department of Civil
  Engineering, 48-415
Massachusetts Institute of
  Technology
Cambridge, MA 02139

Dr. Ronald W. Harvey, US4
U.S. Geological Survey
325 Broadway
Boulder, CO 80303-3328

Dr. Kim F. Hayes
University of Michigan
Department of Civil Engineering
Ann Arbor, MI 48109-2125

Dr. James R. Hunt
Department Civil Engineering
University of California
Berkeley, CA 94704

Dr. Peter R. Jaffe
Department of Civil Engineering
  and Operations Research
Princeton University
Princeton, NJ 08544

Dr. Chad T. Jafvert
Purdue University
School of Civil Engineering
1284 Civil Engineering Building
West Lafayette, IN 47907-1284

Mr. Ray Lappen
Department of Chemical
 Engineering
374 Dow Building
University of Michigan
Ann Arbor, MI 48109

Dr. Richard G. Luthy
Department of Civil Engineering
Carnegie Mellon University
Pittsburgh, PA 15213-3890

Dr. Robert E. Martin
Lyonnaise des Eaux-Dumez
72, Avenue de la Liberte
9200 Nanterre
FRANCE

Dr. John F. McCarthy
Environmental Sciences
 Division
Oak Ridge National Laboratory
P.O. Box 2008
Oak Ridge, TN 37831-6036

Dr. Michael J. McInerney
Department of Botany &
 Microbiology
University of Oklahoma
770 Van Vleet Oval
Norman, OK 73019-0245

Dr. Aaron L. Mills
Department of Environmental
 Sciences
Clark Hall
University of Virginia
Charlottesville, VA 22903

Dr. Huub Rijnaarts
Department of Microbiology
Agricultural University
Hesselink van Suchtelenweg 4
6703 CT Wageningen
THE NETHERLANDS

Dr. M.M. Sharma
Department of Petroleum
 Engineering
University of Texas at Austin
Austin, TX 78712

Professor Chi Tien
411 Link Hall
Department of Chemical
 Engineering
Syracuse University
Syracuse, NY 13244

Dr. T. David Waite
Australian Nuclear Science &
 Technology Organization
Lucas Heights Research
 Laboratories
New Illawarra Road, Lucas
Heights, NSW
Private Mail Bag 1
MENAI NSW 2234
AUSTRALIA

Dr. John C. Westall
Chemistry Department
Oregon State University
Corvallis, OR 97330

Dr. Frank J. Wobber
Program Manager
Subsurface Science Program
Office of Energy Research
  (ER-74)
U.S. Department of Energy
Washington, DC 20545

Dr. John M. Zachara
Earth Sciences Department
Battelle Pacific Northwest
  Laboratory
P.O. Box 999
Richland, WA 99320

CONTENTS

Section I

Mobilizing and Depositing Colloids

Section II

Biocolloid Mobility

Section III

Manipulating Colloids
to Change
Aquifer Permeability

Section IV

Modified Colloids,
Surfactants, and Emulsions

Section V

Abstracts of Poster Presentations

CHAPTER 1

Introduction and Scope

John F. McCarthy and Frank J. Wobber

Attempts to eliminate and remediate contaminant migration cannot succeed if the major pathways and mechanisms of transport are not accounted for. In 1988, the U.S. Department of Energy's Subsurface Science Program established a research program that emphasizes research on the critical, but poorly understood, role of colloidal-size (submicron) particles and biocolloids in facilitating contaminant transport and in remediation. Models of contaminant transport processes typically treat groundwater as a two-phase system in which contaminants partition between immobile solid constituents and the mobile aqueous phase. However, components of the solid phase in the colloidal size range may also be mobile in subsurface environments. Association of contaminants with mobile colloidal particles may, therefore, enhance the transport of strongly sorbing pollutants. Current approaches to monitoring and predicting solute transport generally ignore colloid-facilitated transport mechanisms because little, if any, information is available on the abundance and distribution of colloidal particles in groundwater, their affinity to bind contaminants, or their mobility in subsurface systems. Accurate assessment of current contaminant problems, engineering of containment strategies, and cost-effective remediation approaches are all dependent on a fundamental understanding of transport processes.

The overall goal of DOE research in colloids and biocolloids is to establish a fundamental understanding of the physical-chemical and hydrologic factors that control contaminant mobility in subsoils and groundwater. Particular emphasis is directed at understanding the role of natural organic matter and mobile inorganic particles in the colloidal (submicron) size range in contaminant migration, especially processes controlling the genesis, stabilization, and transport of groundwater colloids. This fundamental understanding will permit the role of mobile colloids to be incorporated in improved contaminant transport model, and will lead to development and evaluation of innovative strategies for in situ manipulation of colloids for waste management and environmental restoration.

It is this last theme that is the focus of this workshop on "Conceptual Strategies to Manipulate Groundwater Colloids for Environmental Restoration." The research question addressed by the workshop is: Can fundamental understanding of processes controlling colloid mobilization and deposition be applied to the development of strategies to manipulate colloid mobility for mitigation or remediation at waste sites?

Fundamental understanding of the chemical and hydrophysical processes controlling colloid mobilization and deposition could contribute to developing novel and cost-effective mitigation and remediation strategies. This research would offer a unique opportunity to apply a fundamental understanding of a contaminant transport mechanism to the development of a practical solution not only to problems of colloid-facilitated contaminant transport, but potentially to development of a generally applicable strategy for in situ isolation or remediation of a number of waste site scenarios.

The objective of this meeting was to examine the current state of science with respect to the feasibility of such approaches, to identify concepts and strategies that in the long term might complement existing engineered cleanup technologies, and to prioritize fundamental gaps in knowledge that must be addressed to develop these capabilities within the next 5 to 20 years. The scope of the meeting included consideration of both organic and inorganic colloids and organic, heavy metal, and radionuclide contaminants, as well as mixed wastes.

The previous Manteo meetings have been a forum to encourage interactions among experts in a broad range of technical specialties. Because the proposed workshop addresses the extension of theoretical understanding of colloid behavior to the complex and heterogeneous realities of natural groundwater systems, we sought the participation of colloid and interface scientists, geochemists, hydrologists, environmental engineers, as well as scientists and engineers working on the more-applied side of hazardous waste management, petroleum engineering and water filtration.

ORGANIZATION OF THE MEETING

The meeting was organized into a series of four sessions focusing on different types of strategies for manipulating colloids. A fifth session considered potential complexities of coupled interactions of geochemical, hydrological, and microbiological processes. In addition to a discussion session at the end of each formal session, participants broke into working groups to reach a consensus on the feasibility of different strategies. The sessions and their objectives are listed here:

SESSION I. Mobilizing and Depositing Colloids
Objective: Can chemical modification of subsurface systems (e.g., pH, redox, and ionic composition) be used to reliably manipulate colloid stability in situ? Will a manipulated system be stable over time?

SESSION II. Biocolloid Mobility
Objective: What controls mobility of microorganisms in porous and fractured media, and can this mobility be manipulated to improve delivery of microbes for bioremediation?

SESSION III. Manipulating Colloids to Change Aquifer Permeability
Objective: Will changes in colloid stability affect the physical structure of a formation to either increase permeability (to enhance bioremediation) or clog pores to create in situ barriers?

SESSION IV. Introducing Modified Colloids, Surfactants and Emulsions
Objective: Surfactants can modify surface binding sites on colloids or can promote formation of colloidal emulsions. Can they be used to control colloid mobility or to increase recovery of sorbed contaminants by pump and treat methods?

SESSION V. Coupled Processes in Colloid Manipulations
Objective: Consolidate previous discussions on how proposed manipulation strategies will be influenced by the multiply-interacting microbiological, chemical, and hydrologic factors in natural systems. Is the cure worse than the disease?

SESSION VI. Working Group Consensus on Manipulation Concepts and Strategies
Objective: Participants divided into working groups to summarize discussions on each topic, develop recommendations on different conceptual strategies, and identify long-term research needs for achieving the most feasible strategies.

The topic of the meeting was broad and interdisciplinary, and specific data directly relating to many of the topics are lacking. Therefore, we relied on moderators to provide an overview of the nature of the problems/approaches and current state of science on each topic. The moderators and discussion leaders were asked to prepare a list of "strawman" strategies to present to the group to both promote and focus discussion. These strawman strategies were presented in the form of tables that included consideration of the key hydrochemical, geochemical, and bio-chemical processes underlying each proposed strategy, the advantages and limitation of the approach, the major gaps in our current knowledge that need to be addressed to make the strategy feasible, and some general rating of the overall feasibility of the strategy. These ideas were further refined in the final session of the workshop. Participants broke into working groups to reach a consensus on the different strategies. At the end of the meeting, the entire group reconvened to arrive at a more general statement of the most promising conceptual strategies for manipulating colloids for environmental restoration (see Section II).

Poster presentations were an important feature of the meeting and was be one of the primary mechanisms for communicating specific scientific studies related to discussions. Posters were available for viewing and discussion at social hours and at breaks and lunches. Extended abstracts of the poster presentations are included at the end of this document.

ORGANIZATION OF THE SUMMARY REPORT

The report is divided into sections summarizing the contents of each session. This summary report begins, however, with a synopsis of the "bottom line" consensus of the workshop on the conceptual strategies judged most feasible. The reports on each session begin with the moderator's overview, followed by the extended abstracts of the speakers within that session. Finally, a summary of the discussion and the results of the deliberations of the discussion groups is provided. The strategies evaluated by each group are summarized in the form of tables presented at the end of each section.

CHAPTER 2

Summary of
Workshop Consensus

John F. McCarthy

The workshop's format allowed the participants to consider a number of specific remediation strategies and scenarios involving manipulation of colloids. Reports from discussion groups focused on topics identified in the different sessions are presented in Session III. At the end of the meeting, all the workshop participants were convened to arrive at an overall consensus on the general strategies that appeared to offer the greatest feasibility. In addition, the gaps in understanding that limit current capability to implement these strategies were identified and prioritized by consensus. Results of these deliberations are summarized in the following.

Three basic remediation scenarios were identified and are outlined in Table 2.1:

- Enhancing contaminant scavenging in pump-and-treat remediation by mobilizing in situ colloids.
- Manipulating in situ colloids and biocolloids to reduce aquifer permeability to reduce contaminant mobility or to isolate a zone of contamination for remedial treatment such as air stripping or soil washing.
- Introduce nonindigenous colloids, specialized biocolloids, or surfactants to optimize contaminant scavenging, reduce permeability of a formation, or enhance bioremediation.

APPROACH 1: ENHANCED CONTAMINANT SCAVENGING IN SUPPORT OF PUMP-AND-TREAT REMEDIATION

Description

Pump-and-treat approaches to aquifer restoration are proven but have limited applicability for removal of contaminants with low solubility in water. Many contaminants of greatest concern, including polychlorinated biphenyls, polycyclic aromatic hydrocarbons, many radionuclides, and toxic metals have high affinities for sorbing to solid-phase sorbents. Given their large surface-to-volume ratio, colloidal particles should act as efficient scavengers of particle-reactive contaminants. Thus, manipulations that would mobilize in situ colloids and permit recovery of them in pumping wells could significantly enhance removal of sorbed contaminants.

Processes

Several strategies for mobilizing colloidal particles in situ were considered in Session I and are described in more detail in the summary of that discussion group. Mobilization can be effected by dissolution of secondary mineral "cementing" phases, by pH manipulations to electrostatically stabilize colloids, by the introduction of specific sorbates to alter the surface charge on the colloids, or by dispersion of colloids by decreasing the ionic strength of the groundwater (see Table 8.1).

Advantages

Contaminant scavenging by mobilized colloids would extend the applications of well-established pump-and-treat techniques to permit recovery of contaminants with low solubility in water.

Limitations

The effectiveness of the manipulations may be limited by difficulties in uniform delivery of the treatment caused by physical heterogeneities in the soil formation. Moreover, the multiphasic nature of natural formations makes it likely that colloids are immobilized by a variety of mechanisms that may react differently to the manipulation. Techniques for maximizing the recovery of mobilized colloids at pumping wells have not been developed.

Gaps in Knowledge

Delivery of the treatment designed to mobilize colloids is poorly understood at the pore level. Uniform delivery of the treatment to zones in which colloid deposits have accumulated within the aquifer is uncertain, and secondary effects of the treatment on other solids is undoubtedly complex. The kinetics of colloid release are poorly understood, as are the dynamics of contaminant redistribution between the immobile solids and the newly mobilized colloids. Finally, efficient recovery of the mobilized colloids and sorbed contaminants is essential or the "remediation" may in fact create even worse problems of contaminant migration by expanding the area affected.

Overall Feasibility

There are a wide variety of approaches available that are likely to be effective, depending on the specific characteristics of a site. Given the potential for spreading contaminants, feasibility studies at uncontaminated sites is recommended.

APPROACH 2: PORE CLOGGING TO REDUCE CONTAMINANT MIGRATION OR TO ISOLATE A FORMATION FOR TREATMENT

Description

This approach involves redistribution of in situ colloids to reduce the permeability of the formation. Several strategies are discussed in more detail in Session III. Colloids must be mobilized and either aggregate to particle sizes that would permit their collection by physical straining at pore restrictions or be mobilized at sufficiently high concentrations to "log-jam" at pore restrictions. An alternative approach involves stimulation of indigenous microorganisms to biofoul/bioplug a formation. The anticipated reduction in aquifer permeability would limit advective transport of the mobile-phase contaminants. Because biofilms are expected to be relatively short-lived as a result of restrictions in the flow of nutrients needed to sustain the bacteria, this strategy may be better suited to the temporary isolation of a limited zone of contamination for treatment by air stripping or soil washing, etc.

Processes

Mobilization of inorganic colloidal particles was discussed in the preceding paragraph. Permeability reductions can occur by physical straining and pore clogging if particles, or aggregates of particles, are of a size comparable to that of the pore restrictions. The processes of mobilization of colloids and subsequent aggregation and collection at pore restrictions must be balanced if the desired effect is to be achieved. Biofilm development involves consideration of cell growth, production of extracellular polymers, substrate supply, and predation (see Tables 4.3.1 and 5.3.1).

Advantages

These in situ strategies offer the potential of reducing migration or enhancing cleanup of zones of contamination that are not accessible to more conventional technologies such as slurry walls or geotechnical barriers. Localized and temporary biofouling may permit treatment without permanently altering the hydrology of the site. Studies with sandstone cores suggest that the strategy would be effective at least for tightly structured formations.

Limitations

In addition to the complexities involved in mobilizing colloids, physical heterogeneities in the formation, especially the presence of preferential flow paths, may limit the extent to which permeability is reduced. Penetration may be limited if pore blockage confines the treatment to a localized area.

Gaps in Knowledge

Predictions and performance assessment of permeability reduction and flow redistribution requires currently unmeasurable information about the pore size distribution of the formation. Although biofouling has been observed around injection wells, a degree of biofouling in a heterogeneous formation sufficient to change regional flow has not been demonstrated.

Overall Feasibility

Permeability reductions resulting from pore clogging by colloids appear to be feasible in tightly structured formations, such as sandstones, but are unlikely to be successful in more-permeable, unconsolidated media. Stimulating biofilm development has promise for isolating localized areas to enhance other remedial treatments.

APPROACH 3: INTRODUCTION OF COLLOIDS, SURFACTANTS, OR BACTERIA FOR CONTAMINANT SCAVENGING, PERMEABILITY REDUCTIONS, OR BIOREMEDIATION

Description

The previous two approaches depend on the presence of indigenous colloids and the ability to mobilize them within the formation. Many of the difficulties and uncertainties involved in these processes may be circumvented by introducing nonindigenous colloids. Introduced colloids can be inorganic particles with known properties suited to the proposed manipulation. Alternative, introduction of surfactants may be useful in scavenging contaminants or in selective alteration of permeability to improve delivery of a treatment to pores (see Table 28.1 and associated discussion). In addition, bioremediation may be significantly improved if microorganisms with specialized degradative capabilities are injected within zones of contamination, especially contamination by refractory compounds (see Table 15.1 and associated discussion).

Processes

Processes controlling the transport and colloidal stability of introduced colloids are relevant to scavenging contaminants, as is redistribution of contaminants sorbed to the aquifer media. Permeability effects beyond a limited zone around the injection well require that the colloids remain stable for some period of time before they aggregate and clog pore restrictions. Similarly, introduced microorganisms must be mobile enough to seed the contaminated zone, but must attach and grow on media surfaces in order for bioremediation to be effective. Relevant processes include colloid stability, aggregation, and deposition. Transport and deposition of bacteria are likely to be controlled by their size and surface properties in ways that should be analogous to processes known to control inanimate particles; however, little research is available to permit predictions of bacterial transport. Furthermore, surface properties and attachment efficiencies of bacteria are altered by growth state and nutrient availability.

Advantages

These strategies are only minimally dependent on the abundance and nature of in situ colloids or on the degradative capabilities of indigenous bacteria because the specific properties of the inorganic colloid, surfactant, or bacterial strain can be selected or engineered to optimize performance at a specific site.

Limitations

Many of the difficulties and uncertainties discussed in the previous strategies apply here. In addition, unanticipated secondary effects of the introduced material may be deleterious.

Gaps in Knowledge

There are gaps in fundamental understanding of colloid behavior in natural systems. Although introduction of "optimized" colloids permits greater control since because behavior and performance can be studied in the laboratory, our capabilities to reliably extrapolate that understanding from the laboratory to the field scales are limited.

Overall Feasibility

Existing understanding and experience suggests that the use of surfactants to improve contaminant scavenging would be feasible and is likely to be successful. Far less is known about the feasibility of introducing colloidal particles either for scavenging contaminants or for permeability reductions. For remediation of recalcitrant contaminants, introduction of specialized bacteria appears to offer potential benefits that appear to justify feasibility studies.

DISCUSSION AND CONCLUSIONS

Because participants had an opportunity to consider most of the issues during the individual discussion sessions, the final phase of the workshop focused on the principal knowledge gaps and research needs that cut across all the proposed manipulation strategies. These are listed here in general order of importance as perceived by the workshop participants.

- *Delivery/transport.* This encompasses consideration of delivery of treatments (reductants, acids, surfactants, nutrients, electron acceptors, etc.), as well as transport of colloids, surfactants, and bacteria to the desired location. In particular, knowledge about delivery and mixing at the pore scale was identified as a major deficiency.
- *Scaling issues.* Understanding of fundamental processes must be able to be transferred from the laboratory scale to the field scale, including inter-

mediate- or pilot-scale feasibility studies. New ways of describing chemical and physical heterogeneity are needed, as is increased understanding of higher-order phenomena expressed at larger scales.

- *Effectiveness of the treatment.* The performance of a proposed strategy must be evaluated relative to the site-specific cleanup needs. Understanding of controlling processes must be generalized in the form of predictive models based on parameters that can be measured or reliably estimated.
- *Permanence/longevity.* For strategies involving deposition or immobilization of colloids (e.g., permeability reductions), the temporal stability of the manipulation must be known. This issue was deemed especially relevant to biofouling.
- *Distribution of contaminants on colloids and aquifer solids.* The mechanisms and kinetics of sorption and desorption of contaminants on colloids and aquifer solids must be known for strategies involving scavenging.
- *Coupled processes.* Any manipulation of the subsurface environment must consider (and may be able to take advantage of) the interactions among geochemical, hydrological, and microbiological processes. Our knowledge, in general, is organized within disciplines, and much more information is needed about how processes interact in natural systems.

TABLE 2.1 Workshop consensus on conceptual strategies for manipulating groundwater colloids for environmental restoration.

PROPOSED MANIPULATION APPROACH	KEY GEO-HYDRO- AND BIOCHEMICAL PROCESSES IN APPROACH	ADVANTAGES/ REASONABLENESS OF APPROACH	LIMITATIONS/ IMPRACTICALITIES IN APPROACH	MAJOR GAPS IN KNOWLEDGE	OVERALL FEASIBILITY
1. Enhance contaminant scavenging in support of pump-and-treat	Mobilization of in situ colloids Dissolution of secondary mineral phases Colloid stabilization Recovery of mobilized colloids	Targeted on contaminants that would be difficult to remove by conventional pump-and-treat In situ remediation applicable to inaccessible zones of contamination	Secondary effects on other solids may be complex Chemical and physical heterogeneities complicate treatment Recovery of mobilized colloids is problematic	Transport/delivery of treatment Kinetics of colloid mobilization Distribution of contaminant on colloids and solids	Several promising approaches, but feasibility must be assessed on site-specific basis
2. Pore clogging to reduce contaminant mobility or to isolate formation for treatment	Mobilization and deposition of in situ colloids and biocolloids Physical straining Aggregation of colloids Blockage of pores Biofilm growth/biofouling Mixing of reactants in situ	In situ remediation applicable to inaccessible zones of contamination Biofouling may permit isolation of treatment zone but be reversible	Physical heterogeneities may reduce effectiveness of blockage Delivery of treatment and mixing at pore scale is uncertain Limited penetration into formation expected for some approaches Low feasibility in higher permeability aquifers	Pore and particle distributions must be known Predicting deposit permeability and flow redistribution Predictive models unavailable Dynamics of biofilm formation and permanence of permeability reduction	Biofouling to isolate small zones for treatment was judged to have a high probability of success Feasibility of other approaches is highly dependent of site characteristics

PROPOSED MANIPULATION APPROACH	KEY GEO-HYDRO- AND BIOCHEMICAL PROCESSES IN APPROACH	ADVANTAGES/ REASONABLENESS OF APPROACH	LIMITATIONS/ IMPRACTICALITIES IN APPROACH	MAJOR GAPS IN KNOWLEDGE	OVERALL FEASIBILITY
3. Introduction of colloids, surfactants, or bacteria for contaminant scavenging, for permeability reduction, or for bioremediation	Transport of "specialized" colloids, surfactants, or bio-colloids to zone of contamination	Less reliance on properties and concentrations of indigenous colloids/bacteria	Physical heterogeneities complicate delivery of colloids	Transport/delivery of colloids and micro-organisms	High chance of success using injected surfactants
	Colloid stability/transport	Permits introduction of specialized colloids or surfactants	Localized deposition around injection area may limit penetration	Redistribution of contaminants on injected particles	Feasibility of injected particles is uncertain due to limited experience
	Contaminant redistribution on colloid/solid phases	Selected or engineered microorganisms can enhance degradation of resistant compounds	Recovery of injected colloids is uncertain	Predictive capabilities are limited by physical heterogeneities and inadequate theory	Potential benefits of delivering specialized microorganisms merit feasibility studies at field sites
			Complex coupled processes control delivery, growth and ecology of introduced microorganisms		

SECTION I

MOBILIZING AND DEPOSITING COLLOIDS

CHAPTER 3

Overview

John C. Westall and Philip M. Gschwend

Colloidal dispersions consist of finely divided particles of one phase uniformly dispersed in another phase. Although these two phases may be any combination of solids, liquids, and gases, the colloidal systems of primary interest in ground waters are dispersions of solids in liquids, also referred to as sols.

Colloidal particles bridge the gap between true dissolved solutes and true bulk phases: colloidal particles are larger than typical solutes in true solution and are smaller than most phases with easily recognizable bulk properties. In developing physical–chemical models of colloidal systems, we invoke concepts that pertain both to particles and to bulk phases. The common feature of colloidal systems is interactions over the distance scale of 1 to 1000 nm. Thus, although we focus on colloidal dispersions of solids in liquids, other colloidal systems such as films, gels, foams, micelles, emulsions, and microemulsions have critical dimensions of the same magnitude and are handled with many of the same theoretical and experimental techniques. These other colloidal systems are of interest as we consider methods for environmental restoration. Recent books by Everett (1988) and Goodwin (1982) provide thorough, easily readable introductions into the fascinating world of colloid science.

The fundamental question with regard to the presence of colloidal particles in groundwater is colloid stability; that is, when do the particles remain dispersed (*a stable colloid*) and when do they settle out (*an unstable colloid*)?

The simple answer to this question is that small particles tend to remain in suspension and large particles tend to settle out. Colloidal particles that are smaller than approximately 1 to 10 µm are maintained in suspension by Brownian motion; these particles would remain in suspension indefinitely if collisions did not result in (1) aggregation of the small particles to form larger particles or (2) adhesion of the colloidal particles to surfaces of large particles. Particles that are larger than approximately 1 to 10 μm are prone to settle out of suspension eventually if not resuspended by fluid flow. Thus the question of colloid stability boils down to two issues: (1) the frequency of collisions between particles and (2) the probability of particles adhering if they collide. In this overview, we shall discuss the first step briefly and the second step in more detail.

Collision frequency. O'Melia (1989) has reviewed particle-particle interactions in aquatic systems and reached the conclusion that models for collision frequency are generally satisfactory. Thus, if all collisions result in adhesion, the models are satisfactory. However, if adhesion does not occur with every collision, the models generally underestimate the rate of adhesion. More information on processes determining collision frequency is available in the Overview to Session III.

Probability of adhesion. The starting point for the analysis of colloid stability is usually the DLVO (Derjagin-Landau-Verwey-Overbeek) theory. The attractive forces and repulsive forces between two particles are calculated as a function of particle-particle separation distance. The curves of potential energy versus particle-particle separation distance that result from these calculations typically show an energy barrier that must be overcome for colliding particles to adhere. The primary attractive forces are dispersion forces, and the primary repulsive forces are long-range electrical forces and short-range Born repulsion.

The attractive energy is primarily the result of London-van der Waals dispersion forces and is related to the polarizability of the particles and the distance of the separation between the particles. For colloidal particles, the attractive energy declines with separation according to

$$V_A = -A_H / d^n$$

where V_A is the energy of attraction, A_H is the Hamaker constant, which depends on the difference in polarizability between the dispersed particles and the dispersing medium, d is the separation, and n is an exponent, the value of which is $1 < n < 3$, depending on the geometry of the system (flat-flat, flat-sphere, sphere-sphere, etc.). If a colloidal particle and the dispersing solvent are of similar polarizability, then the tendency for the particles to seek out one another and to separate from the solvent is small (i.e., there is little tendency for the particles to adhere).

The electrical repulsion energy is due to the repulsion of like-charged surfaces. Formally, three paradigms are recognized for treating surface charge: (1) constant charge surfaces (e.g., clays); (2) constant potential surfaces (e.g., AgI); and (3) constant chemical affinity surfaces (e.g., oxides). These three components of surface charge result from specific chemical interactions or specific adsorption.

Another component of surface charge is the counter charge that is held electrostatically to balance the specific-adsorption charge. The counter charge is usually treated through an electrical double layer model based on the Gouy-Chapman-Stern theory. According to this theory, the repulsive energy drops off exponentially with distance from the surface according to

$$V_R \propto \exp(-\kappa d)$$

where κ = the inverse of the characteristic width of EDL, which is proportional to the square root of the ionic strength for a z:z electrolyte, $\kappa \propto I^{\frac{1}{2}}$. The characteristic thickness of this layer, κ^{-1}, as a function of ionic strength, is: $I = 1$ mM, $\kappa^{-1} = 10$ nm; $I = 10$ mM, $\kappa^{-1} = 3$ nm; $I = 100$ mM, $\kappa^{-1} = 1$ nm.

The following example illustrates the effect of adding a specifically adsorbing ionic solute to a colloidal suspension, in which the charge on the particles is opposite to the charge of the specifically adsorbing solute:

- At "low" concentrations of solute, colloidal surface charge is not significantly affected, and electrostatic repulsion between particles is still strong. Under these conditions, the suspensions remain stable.
- At "mid" concentrations of solute, specific adsorption tends to neutralize the inherent colloidal surface charge. Under these conditions, particle-particle collisions result in adhesion and coagulation.
- At "high" concentrations of solute, the colloidal surface charge may reverse, restabilizing the particles as electrostatic repulsions return.

Another repulsive interaction that has not yet been widely documented in environmental systems is steric repulsion. This repulsion occurs when particles that are coated with large polymers collide and the polymeric coatings overlap. Factors affecting the extent of steric repulsion are the similarity of the coating to the solvent and to the particle, and the expansion or contraction of the polymer according to the properties of the solvent.

At very small distances between colloidal particles (i.e., of order nanometers), still other processes affect the interaction between colliding colloidal particles. For example, the solvent may be structured near surfaces and surface moieties may be hydrated. Under such circumstances, it may be necessary to displace structured water and/or dehydrate surface sites to enable physical contact between two colloidal particles. Such solvent effects may offer another barrier to adhesion, especially at very short particle-particle spacings.

The state of our knowledge on colloid stability varies widely. Models of physics of Brownian particles and their frequencies of collision appear to be satisfactory. Models of the *chemistry* limiting colloid-colloid attachment (DLVO theory) appear satisfactory for cases where they predict collision efficiencies of >0.001 (i.e., more than 1 in 1000 collisions result in particle attachments). However, at lower predicted collision efficiencies, experimental observations are not well-understood. Also, when specific adsorption phenomena must be included to predict collision efficiencies, theories do not match data well. Other theoretical issues remain unmastered, including the role of steric factors in colloidal stability, and approaches to dealing with microscale heterogeneity of solid surfaces. Finally, and possibly most limiting, is our ability to transfer theoretical understanding or even laboratory findings to make predictions in field settings.

REFERENCES

Everett, D.H., Basic Principles of Colloid Science. Royal Society of Chemistry, London, 1988.

Goodwin, J.W. (ed.), Colloidal Dispersions. Royal Society of Chemistry, London, 1982.

O'Melia, C.R., Particle-particle interactions in aquatic systems. *Colloids Surfaces* 39:255–71, 1989.

CHAPTER 4

Particle Deposition in Porous Media in the Presence of a Repulsive Double Layer Force

Chi Tien

The transport and deposition of particles associated with the flow of suspensions through porous media can be analyzed with the use of trajectory calculations based on appropriate porous media models (Tien 1989). For the case in which there is insignificant repulsive double-layer force between the particles and surfaces of the solid matrix of the porous medium, theoretical calculations give results that are in reasonably good agreement with experiments (Payatakes et al. 1974; Rajagopalan and Tien 1976). On the other hand, if there is a repulsive double-layer force barrier between the particles and the medium surface, theoretical calculations fail to agree with experiments even qualitatively (Rajagopalan and Tien 1977; Onorato 1980).

A number of hypothesis have been advanced to explain this disparity between theory and experimental data concerning particle deposition. Corresponding to these hypotheses, modifications of trajectory analysis have been attempted to eliminate or reduce the observed disparity. These modifications include a formulation of new models of porous media (Vaidyanathan et al. 1989) and considerations of the effect of the roughness of the media surfaces (Vaidyanathan and Tien 1988) and the nonuniformity of particle and media surface potentials (Vaidyanathan and Tien 1989). Nevertheless, results obtained by incorporating these modifications, in general, were in poor agreement with experiments, although improvement was obtained in certain cases.

A patched surface model is proposed for analyzing particle deposition in the presence of a repulsive double-layer force. The surface on which deposition may take place is assumed to have a number of discrete patches composed of ionizable sites. Justification of this assumption is based on the fact that oxide surfaces (silicates in particular) often indicate heterogeneity in the types of surface groups and that oxides are the most likely substances encountered in particle deposition. A number of investigators (e.g., Armistead et al. 1969; Goates and Anderson 1956; Zhdanov et al. 1987; Mathias and Wannemacher 1988) have shown that the hydroxyl groups on silicate surfaces are not all identical, and their dissociative tendencies may differ. The fraction of the more active sites was found to vary from 18 to 30%. The specific assumptions used in formulating the model are

1. Active sites are present in patches.
2. Only the more active sites become charged when placed in an aqueous environment.
3. The patches are circular in shape and identical in size. The size is much smaller than the characteristic dimension of the surface but much greater than the size of the individual site.
4. The distance between adjacent sites are sufficiently large to be considered independent of one another.

Based on these assumptions, the electric double-layer repulsive force between a charged surface and particle arises from the interactions between the particle and the patches. These interactions lead to formation of certain excluded areas on the surface, upon which deposition cannot take place. For a single patch, the extent of the exclusion may coincide with the size of the patch, depending on the surface charges (or potentials) as shown in Figure 1.

The excluded area of deposition can be determined from data relating to the surface interaction force between a particle and a surface path as shown in Figure 1. Consequently, a method for estimating the double-layer force between a particle and a patched surface is required. Such a method was recently proposed by Vaidyanathan (1989).

Using the method of Vaidyanathan, together with the equations of particle motion, the extent of exclusion expressed as $\Omega + l_c$, where Ω is the radius of the patch and l_c is either a positive or negative quantity for particles of given size and with specified surface charge (or potential). The filter coefficient may be expressed as

$$0 \text{ if } (f)\left(\frac{\Omega c}{\Omega}\right)^2 \geq 1 \frac{\lambda}{(\lambda_o)_f} = 1 - f\left(\frac{\Omega + c}{\Omega}\right)^2 \text{ if } \left(\Omega + \frac{c}{\Omega}\right) < 1$$

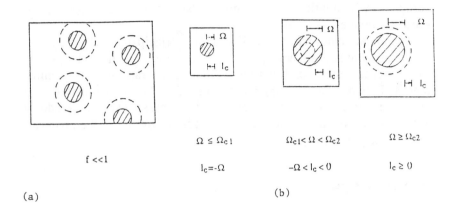

$\Omega \leq \Omega_{c1}$ $\Omega_{c1} < \Omega < \Omega_{c2}$ $\Omega \geq \Omega_{c2}$

$f \ll 1$

$l_c = -\Omega$ $-\Omega < l_c < 0$ $l_c \geq 0$

(a)   (b)

Figure 1. Schematic representations of a patch surface (a) and the presence of excluded area (b).

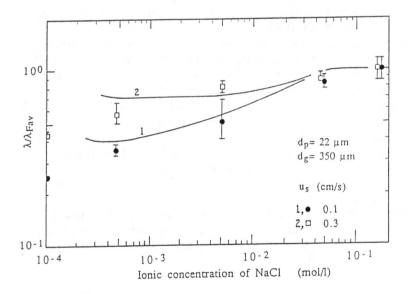

Figure 2. Comparison between model predictions and experiments. d_p and d_g are particle and filter grain diameters, respectively. Data reported by Vaidyanathan (1989).

where λ_o is the filter coefficient and $(\lambda_o)_f$ is the filter coefficient if there is no repulsive force barrier.

A large number of sample calculations were made and the results were compared with available experimental data. A typical comparison is shown in Figure 2.

The model is not without its limitations; principally, the presence of two adjustable model parameters, f and l_c. However, from the sample calculations, it was found that these adjustable parameters have limited ranges. In particular, by using f = 0.2 and Ω = 0.5 μm, reasonably good agreements were obtained in most if not all cases. Thus, the model is indeed predictive and has an accuracy that is significantly better than earlier models.

REFERENCES

Armistead, C.G., Tyler, A.J., Hambleton, F.H. Mitchell, S.A., and Hockey, J.A., The surface hydroxylation of silicas, *J. Phys. Chem.*, 73:3947, 1969.

Goates, J.R. and Anderson, K., Acidic properties of quartz. *Soil Sci.*, 81:277, 1956.

Mathias, J. and Wannemacher, G., Basic characteristics and applications of aerosol, *J. Colloid Interface Sci.*, 125:61, 1988.

Onorato, F.J., The effect of surface interactions on past deposition, *Chem. Eng. Commun.*, 7:363, 1980.

Payatakes, A.C., Tien, C., and Turian, R.M., Trajectory calculation of particle deposition in deep bed filters: Parts I and II, *AIChE J.*, 20:889, 1974.

Rajagopalan, R. and Tien, C., Trajectory analysis of deep-bed filtration with sphere-in cell model, *AIChE J.*, 55:256, 1976.

Tien, C., *Granular Filtration of Aerosols and Hydrosols*, Butterworths, 1989.

Vaidyanathan, R. and Tien, C., Hydrosol deposition in granular beds, *Chem. Eng. Sci.*, 43:289, 1988.

Vaidyanathan, R. and Tien, C., Hydrosol deposition in granular beds - an experimental study, *Chem. Eng. Commun.*, 81:123, 1989.

Vaidyanathan, R., Yao, C., and Tien, C., Flow effects in granular filtration, *Chinese Inst. Chem. Eng.*, 20:209, 1989.

Vaidyanathan, R., Colloid deposition in granular media under unfavorable surface condition. Ph.D. Dissertation, Syracuse University, Syracuse, NY, 1989.

Zhdanov, S.P., Kosheleva, L.S., and Titova, T.I., IR study of hydroxylated silica, *Langumir*, 3:960, 1987.

CHAPTER 5

Aggregation Kinetics and Fractal Structure of Colloidal Oxides

T.D. Waite, R. Amal, and J.A. Raper

Aggregate size and rate of change in aggregate size are expected to be important determinants of the trapping efficiency of colloidal particles by porous media (O'Melia 1985). However, the inability to quickly measure the size distribution of colloidal aggregates or to predict the change in size with change in solution conditions, coupled with the inability of models based on Derjaguin-Landau-Verwey-Overbeek (DLVO) theory to accurately predict the dependency of capture of sub-micron particles on solution conditions (Elimelech and O'Melia 1990), renders prediction of capture efficiency in this size range difficult.

The study of colloidal aggregation phenomena has been made easier in recent years with the advent of a variety of instruments based on dynamic light scattering (DLS), or photon correlation spectrometric (PCS) principles (Chu 1974; Cummings and Pusey 1977; Pecora 1983). Dynamic light scattering methods have been used in the current work to measure the change in particle size distribution with time and hence to study the aggregation kinetics of hematite particles. In previous work in which DLS was used for the measurement of particle size in aggregating systems, consideration was given only to the initial stages of aggregation in which single particles combine to form doublets (Novich and Ring 1985). In this work, however, long times are considered so that multiplets are also formed. The aggregation results obtained from the light scattering experiments are compared with the predictions of a theoretically derived model in which Smoluchowski's kinetic equation has been modified using a rate constant determined from the repulsion and attraction energies, and account is taken of the nature of packing within the aggregate structure using fractal parameters determined by static light scattering (SLS) techniques (Martin 1986).

Consideration is given to the effect of suspension conditions (pH, temperature, particle number, and electrolyte concentration) on aggregation kinetics and aggregate structure. In addition, the effects of addition of a well-characterized fulvic acid on aggregation kinetics and aggregate structure are investigated. Complete details of these studies are reported elsewhere (Amal et al. 1990a, b and unpublished data).

The agglomerate growth process can be described by a series of 'reactions' of the type

$$n_i + n_j \rightarrow n_{i+j} \tag{1}$$

where n_i, n_j, and n_{i+j} are the number concentration of the 'reactants' and 'product,' respectively. Following the analogy to chemical reactions, a rate constant k_{ij} for the formation of aggregates can be defined as

$$-\frac{dn_i}{dt} = -\frac{dn_j}{dt} = k_{ij} n_i n_j \tag{2}$$

There is also a rate of 'consumption' of aggregates l as 'reactions' to form aggregates of higher order are occurring simultaneously. The balance of these 'reactions' is the classical Smoluchowski kinetic equation (Sheludko 1966)

$$\frac{dn_l}{dt} = \frac{1}{2} \sum_{i=1, j=l-i}^{l-1} k_{ij} n_i n_j - n_l \sum_{i=1}^{\infty} k_{il} n_i \tag{3}$$

Eq. (3) can be numerically integrated to give the variation of the number of aggregates l with time, provided that a suitable expression for k_{ij} (and k_{il}) is used.

If the particles are indifferent to each other and sufficiently small to have negligible settling velocity, the only mechanism promoting collision between them is Brownian motion. In this case, the rate constant can be based on Fick's first law with the coefficient of diffusivity estimated from the Stokes-Einstein equation. The result can be written as (Sheludko 1966)

$$k_{ij}(r) = \frac{2k_BT}{3\eta} \left(\frac{1}{r_i} + \frac{1}{r_j} \right) (r_i + r_j) \qquad (4)$$

where r_i, r_j are the radii of the aggregates.

In many practical cases, however, the rate of aggregation is retarded by the presence of repulsive forces between the particles, and the rate given by Eq. (4) grossly overestimates the aggregates growth.

To account for the retardation in the rate of aggregation, a correction factor, known as the stability ratio, is defined as

$$W_{ij} = \frac{k_{ij(r)}}{k_{ij}} \qquad (5)$$

where $k_{ij(r)}$ is given by Eq. (4) and k_{ij} is the corrected rate constant to be used in Eq. (5). The stability ratio can be determined by an expression derived by Füchs (1934), which can be written as

$$W_{ij} = (r_i + r_j) \int_{r_i + r_j}^{\infty} \frac{exp\left(\dfrac{V_T(r)}{k_BT} \right)}{r^2} \, dr \qquad (6)$$

where r is the distance between the center of aggregates and $V_T(r)$ the total interactive energy between the pair of aggregates. The total interactive energy is given by

$$V_T(r) = V_R(r) + V_A(r) \qquad (7)$$

where $V_R(r)$ is the repulsive and $V_A(r)$ the attractive energy of interaction, respectively. The repulsive energy of interaction has been determined in this work using the expression of Hogg et al. (1966) and the attractive energy has been attributed to Van der Waal's forces. Appropriate expressions are given by Amal et al. (1990).

For the complete prediction of aggregate growth with time, it is necessary to attribute a size to the assembly of particles that constitutes each aggregate. In this work, a fractal geometry for the growing aggregate was assumed. In this case, by definition, an aggregate of mass M contained within a radius R about its center can be described by (Weitz et al. 1985):

$$M \propto R^{d_f} \tag{8}$$

where d_F is the fractal dimension of the aggregate of value < 3. If the primary particles remain intact upon aggregation, then the mass of an aggregate is proportional to the number of the primary particles n within it. Hence, the spherical radius of an aggregate composed of n particles is

$$r_n = n^{\frac{1}{d_F}} r_1 \tag{9}$$

The theoretical average size of the aggregates is then estimated using the mass mean size equation

$$\bar{r} = \frac{\sum f_i r_i^4}{\sum f_i r_i^3} \tag{10}$$

where f_i is the number distribution of aggregates with i particles.

Static light scattering experiments give access to the position correlations between particles in the aggregate and consequently appear to be a very useful tool to measure the fractal dimension (Jullien and Botet 1987). In a scattering experiment, a beam of light is directed on to a sample and the scattered intensity (photon counts) is measured as a function of an angle θ to the incident direction. The incident and scattered beam are characterized by the momentum transfer Q, where

$$Q = |Q| = \frac{4\pi n \sin(\frac{\theta}{2})}{\lambda_o} \tag{11}$$

where λ_o is the wavelength in vacuum, and λ is refractive index of the medium.

The intensity scattered (I_M) from a single fractal aggregate of mass M can be related to two factors that describe the aggregate structure (Teixera 1988),

$$I_M \sim M^2 P(Q) S(Q) \qquad (12)$$

where P(Q) is the form factor and S(Q) the interparticle structure factor.

The form factor is related to the shape of the particle and the contrast (the difference between the scattering properties of the particles and the solvent), whereas the interparticle structure factor S(Q) takes into account interparticle correlations and describes the spatial arrangement of the particles. At Q small compared to $1/r_0$ (where r_0 is the primary particle radius) but large compared to 1/R, where R is the aggregate radius, i.e.,

$$\frac{1}{R} < Q < \frac{1}{r_o}$$

P(Q) is independent of Q; hence I(Q) is proportional only to its structure factor. Under these conditions, it can be shown that (Chen and Teixera 1986)

$$S(Q) \qquad Q^{-d_F} \qquad (13)$$

provided R is much larger than r_0 and provides a relatively simple means of obtaining the fractal dimension of a colloidal aggregate.

Samples of uniformly sized, approximately spherical hematite particles were prepared by forced hydrolysis of an homogeneous iron salt solution under strictly controlled conditions, as described by Matijevic and Scheiner (1978) and slightly modified by Penners and Koopal (1986). Hematite samples were prepared and diluted to $1.95 \times 10^{-4} M$. This concentration corresponds to 2.25×10^{13} particles per liter, assuming the presence of singlets only.

The particle size distributions of the aggregating particles were measured with a Malvern 4700 PCS system, utilizing a 15 mW, 633 nm He-Ne laser. Both Rayleigh-Gans and Mie theory were used to relate the intensity of the light scattered by each size population to its mass. Both procedures were found to give comparable results. The Malvern 4700 was also used to undertake static light scattering studies. The zeta potential of the particles was measured with the Malvern Zetasizer IIc system, which uses a 5 mW, 633nm He-Ne laser as light source.

Aggregation and scattering experiments were conducted at a zeta potential of 48 ± 2 mV (corresponding to a suspension of pH 3) at temperatures of 25, 35, and 55°C, and salt (KCl) concentrations of 50, 60, and 80 mM. For each temperature and salt concentration, the variation in mean hydrodynamic diameter with time was determined by dynamic light scattering measurements every 2 to 3 min for up to 1 h. The intensity (photon counts) of scattered light was measured at angles ranging from 15° to 90° and for 5 s at each angle. To comply with conditions governing the relation of scattering exponent and fractal dimension (discussed in the following), these static light scattering experiments were carried out when the aggregates, containing a large number of primary particles, were typically about 1 μm in diameter.

Studies of aggregation kinetics and aggregate structure were also performed after adsorbing varying amounts of a naturally occurring organic acid (Suwannee River fulvic acid). Hematite suspensions were mixed with different concentrations of fulvic acid for at least 24 h before each aggregation experiment was performed. (Additional details of these studies are described by Amal et al. 1991 unpublished data).

The effects of ionic strength and temperature on variation in aggregate size with time are shown in Figures 1a and b. At 50 mM KCl, the repulsion barrier is still significant; thus, the aggregation rate increases markedly with temperature because it affects both the repulsive energy as well as the rate of particle diffusion. Previous modelling of the effect of temperature on the aggregation process indicated close correspondence between experiment and theory, with the major influence of temperature being on the repulsive energy of interaction and, hence, on the stability ratio (Amal et al. 1989). Indeed, for 50 mM KCl, the stability ratio is on the order of 10^3 at 25°C and decreases to 10^2 and 1 for increases in temperature to 35 and 55°C, respectively. The change in aggregation rate attributable to temperature dependence of diffusion was found to be minor. At 80 mM KCl, attractive van der Waal forces dominate and the temperature affects diffusion only with a resultant relatively insignificant effect on aggregation rate.

The slopes of the logarithmic plots of scattered intensity (photon count) as a function of angle for 50, 60 and 80 mM KCl at different temperatures are given in Table 1 and represent the fractal dimensions of the hematite aggregates. The fractal dimensions obtained vary with aggregation mechanism. For example at 25°C, for slow ("reaction limited") aggregation (50 mM KCl) the fractal dimension d_F is 2.86 ± 0.05; for aggregation

Table 1. Scattering exponents ($\pm 1\sigma$) obtained from static light studies of hematite at different ionic strengths and temperatures.

Temperature	KCl Concentration		
	50mM	60mM	80mM
25°C	2.86±0.05	2.55±0.06	2.33±0.06
35°C	2.56±0.06	2.36±0.04	2.30±0.07
55°C	2.37±0.07	2.32±0.04	2.31±0.03

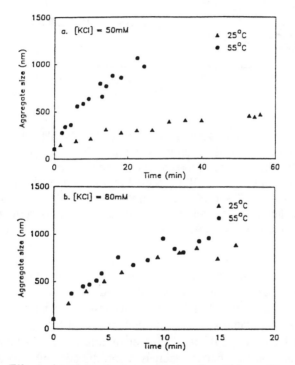

Figure 1. Effect of temperature on aggregation of hematite particles (a) under reaction limited conditions ([KC1]=50mM), and (b) under diffusion limited conditions ([KC1]=80mM).

approaching the diffusion limited regime (60 mM), d_F is 2.55 ± 0.06; and for diffusion limited aggregation (80 mM), d_F is 2.33 ± 0.06.

At the slow rate of aggregation in which the probability of sticking is low, the aggregating clusters will have the opportunity to explore a large number of possible mutual configurations, which leads to some interpenetration and, therefore, denser aggregates. In contrast, in diffusion-limited

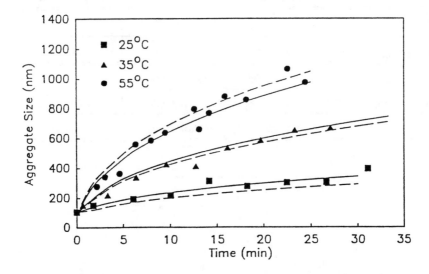

Figure 2. Comparison of aggregation kinetics for hematite at 25°C, 35°C, and 55°C as determined by dynamic light scattering with kinetics predicted using the modified Smoluchowski model described by Amal et al. (1990a). Model predictions in which an equivalence of R_h and R_g is assumed are presented (---) as are predictions incorporating R_h/R_g ratios determined from static light scattering data (see Amal et al., 1990b).

aggregation, the interior of the clusters is effectively screened from penetration because approaching particles readily adhere to other particles (high sticking probability), resulting in a more tenuous structure.

At higher temperatures, the fractal dimensions of the aggregates in 50 and 60 mM KCl are significantly lower than at 25°C. As temperature increases, the repulsion barrier decreases and the sticking probability increases, resulting in much less penetration to the interior of the clusters with resultant formation of more tenuous aggregates. As expected in the absence of interparticle significant (or interaggregate) repulsion, temperature has little effect on the fractal dimension of aggregates formed under diffusion-limited conditions (80 mM KCl) with values of d_F at 25, 35, and 55°C are all about 2.3. Similar effects of temperature on aggregate structure have recently been reported by Tang et al. (1988) in studies of fractal structure of silica aggregates.

The fractal dimension of 2.3 obtained here under diffusion-limited aggregation conditions (no repulsion barrier) is only slightly less than the value of 2.53 expected from theoretical considerations for aggregates growing via particle-cluster diffusion limited aggregation (DLA) (Witten and Sander 1981; Meakin 1983). In DLA, clusters grow purely from monomers, which approach the cluster with a random-walk trajectory. The random-walk nature of the approaching monomer favors growth on the extremities of the cluster, and thus, open, ramified geometries develop. Cluster-cluster aggregation (CA) is a variation of DLA in which clusters grow from existing clusters as well as from monomers. In CA, d_F is reduced substantially because two fractals are extremely unlikely to penetrate without contact. Widely spread, loosely packed structures are produced in cluster-cluster aggregation. Fractal dimensions on the order of 1.75 are to be expected where cluster-cluster aggregation is the dominant mechanism (Meakin 1985; Kolb et al. 1983). The fractal dimensions of $d_F > 2.3$ obtained here are consistent with reaction-limited particle-cluster aggregation. These values are generally considerably higher than results reported by other workers for aggregates of silica and gold for which cluster-cluster aggregation appears to be a more appropriate model (Schaefer et al. 1984; Weitz et al. 1984; Wilcoxon et al. 1988). As can be seen from Figure 2, good correspondence between theory and experiment is obtained when these values of d_F are used to obtain aggregate size estimates using the modified Smoluchowski model. This result is particularly satisfying since the model predictions are obtained quite independently of the experimental data (i.e. no fitting parameters are required).

Adsorption of fulvic acid to hematite surfaces significantly alters the surface potential, and thus dramatically modifies the dependency of kinetics of aggregation on electrolyte concentration. At high surface coverage (fulvic acid concentrations in the range 3 to 45 mg/L) and as found in the organic-free case, the fractal dimension is dependent on the rate of aggregation (as controlled by salt concentration). At low salt concentrations (<50 mM KCl) (reaction-limited conditions), the fractal dimensions (2.4 to 2.7) are similar to those obtained in the organic-free case (2.3 to 2.8), but at salt concentrations high enough to induce diffusion limited aggregation, significantly lower fractal dimensions are observed (2.0 to 2.1) suggesting the occurrence of some flocculation bridging. At low surface coverage (1.0 and 2.5 mg fulvic acid/L), significantly higher fractal dimensions (i.e., more tightly packed structures) than in the organic-free case are obtained and extensive restructuring (as evidenced by increasing fractal dimensions) occurs over time. These observations at low fulvic acid concentrations are associated with partial surface coverage by organic molecules resulting in the presence of both positively and negatively charged groups on the hematite surface. These groups of opposite charge apparently induce tight interparticle association.

REFERENCES

Amal, R., Coury, J.R., Raper, J.A., Walsh, W.P., and Waite, T.D., *Colloids and Surfaces,* 46:1, 1990a.

Amal, R., Coury, J.A., Raper, J.A., and Waite, T.D., in *Proceedings of 5th International Symposium on Agglomeration,* The Institution of Chemical Engineers, Rugby, 1989, 537.

Amal, R., Raper, J.A., and Waite, T.D., *J. Colloid Interface Sci.,* 140:158, 1990b.

Chen, S.H. and Teixera, J., *Phys. Rev. Lett.,* 57:2583, 1986.

Chu, B., *Laser Light Scattering,* Academic Press, New York, 1974.

Cummings H.Z. and Pusey, P.N., in H.Z. Cummings and E.R. Pike (eds.), Photon Correlation Spectroscopy and Velocimetry, Plenum, New York, 1977, pp. 164-99.

Elimelech, M. and O'Melia, C.R., *Environ. Sci. Technol.,* 24:1528-36, 1990.

Füchs, N., *Z. Phys.,* 89:736, 1934.

Hogg, R., Healy, T.W., and Fuerstenau, D.W., *Trans. Faraday Soc.,* 62:1638, 1966.

Jullien, R. and Botet, R., *Aggregation and Fractal Aggregates,* World Scientific, Singapore, 1987.

Kolb, M., Botet, R., and Jullien, J., *Phys. Rev. Lett.,* 51:1123, 1983.

Martin, J.E., *J. Appl. Cryst.,* 19:25, 1986.

Matijevic, E. and Scheiner, P., *J. Colloid Interface Sci.,* 63:509, 1978.

Meakin, P., *Phys. Rev. A,* 27:1495, 1983.

Novich, B.E. and Ring, T.A., *J. Chem. Soc. Faraday Trans.,* 81:1455, 1985.

O'Melia, C.R., *ASCE J. Environ. Eng.,* 10:167, 1985.

Pecora, R., in *Measurement of Suspended Particles by Quasi-Elastic Light Scattering,* Dahneke, B.E., Ed., Wiley, New York, 1983, 3.

Penners, N.H.G. and Koopal, L.K., *Colloids Surf.,* 19:337, 1986.

Schaefer, D.W., Martin, J.E., Wiltzius, P., and Cannel, D.S., *Phys. Rev. Lett.,* 52:2371, 1984.

Sheludko, A., *Colloid Chemistry,* Elsevier, 1966.

Tang, P., Colflesh, D.E., and Chu, B., *J. Colloid Interface Sci.,* 126:304, 1988.

Teixera, J., *J. Appl. Cryst.,* 21:781, 1988.

Weitz, D.A., Huang, J.S., Lin, M.Y., and Sung, J., *Phys. Rev. Lett.,* 54:1416, 1984.

Weitz, D.A., Lin, M.Y., and Sandroff, C.J., *Surf. Sci.,* 158:147, 1985.

Wilcoxon, J.P., Martin, J.E., and Schaefer, D.W., *Phys. Rev. A* 39:2675, 1988.

Witten, T.A. and Sander, L.M., *Phys. Rev. Lett.,* 47:1400, 1981.

CHAPTER 6

Mobility of Natural Organic Matter Injected Into a Sandy Aquifer

John F. McCarthy, Liyuan Liang,
Philip M. Jardine, and Thomas M. Williams

The transport of natural organic matter (NOM) in groundwater is being studied to improve capabilities to predict the subsurface transport of contaminants that sorb to mobile NOM. The research objectives are (1) to determine if NOM is mobile in subsurface environments and (2) to elucidate the chemical and hydrological properties of aquifers that influence transport of NOM and inorganic colloids. In the summer of 1990, a field experiment involving injection of natural NOM was conducted in a shallow, sandy, coastal plain aquifer at Hobcaw Field within the Baruch Forest Science Institute of Clemson University located in Georgetown, South Carolina. The aquifer media was approximately 90 to 95% sand, with 3 to 10% clay, 0.3 to 4.1 mg/g total iron, and 0.03 to 0.05% organic carbon. The primary objective of the study was to determine the field scale mobility of NOM by injecting 80,000 L of NOM-rich "brown water" (66 mg carbon/L) from a nearby terrestrial stream; concomitant effects of the injection on inorganic colloids and on the abundance of groundwater bacteria were also monitored. The injection involved a forced-gradient dipole well array with injection and pumping (withdrawal) wells separated by 5 m and multilevel sampling wells at 1.5 and 3 m from the injection well (Figure 1). Multilevel peizometers were also installed adjacent to each set of wells to monitor hydraulic gradients. The forced gradient was established by recirculating water from the pumping well to the injection well at a flow rate of 4 L/min. After several weeks of recirculation, the NOM solution was injected at 4 L/min over 13 d, whereas the withdrawal water (also 4 L/min) was discarded. A nonreactive tracer (Cl^-) was injected with the first 4000 L of NOM solution to provide information on the average pore water velocity and dispersion characteristics of the media.

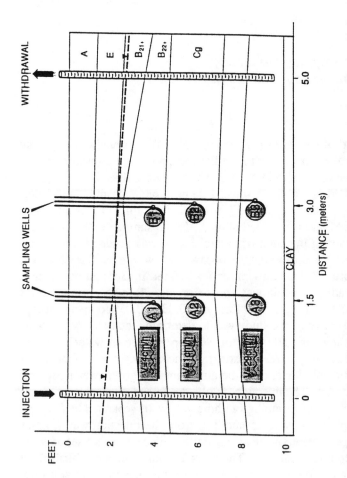

Figure 1. Cross-sectional diagram of the forced gradient dipole well array installed at Hobcaw Field for the NOM injection experiment conducted in June–November 1990. Average pore water velocity of the three sampling horizons is indicated. Soil type for the different horizons are indicated at the right.

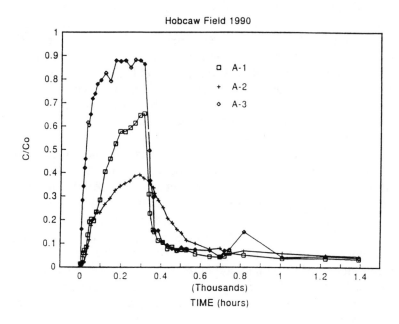

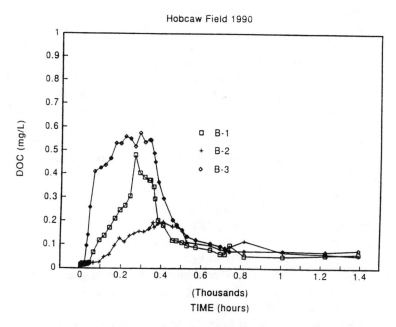

Figure 2. Concentration history of total organic carbon concentrations (expressed as reduced concentration, C/C_o) in groundwater collected from the (A) A-wells and (B) B-wells for 1400-h following initiation of the NOM injection.

The NOM moved rapidly through the aquifer (Figure 2), with an apparent retardation that agreed fairly well with that predicted based on laboratory studies using NOM and aquifer material from the site. The NOM breakthrough did show many features predicted from laboratory studies.

(1) In both laboratory columns and in the field, approximately 10% of the NOM was rapidly transported, whereas the remaining NOM was retarded. Modeling of the NOM transport in the laboratory suggests this initial breakthrough is the result of slow adsorption kinetics for a component of the NOM (Jardine et al. 1989; 1991).

(2) Although some of the NOM was not retarded, there was no evidence of size exclusion in either the laboratory columns (Dunnivant et al. 1991) or in the field. ("Size exclusion" refers to the breakthrough of the NOM before that of the nonreactive tracer resulting from exclusion of the larger organic macromolecules from the smaller pores).

(3) Extensive tailing of NOM breakthrough curves was observed both in laboratory columns (Dunnivant et al. 1991) and in the field. Based on modeling studies of NOM transport in laboratory columns, the tailing may be attributable to nonequilibrium reactions, isotherm nonlinearity, and the differences in adsorption processes for the multiple subcomponent solutes that comprise natural NOM.

(4) Both laboratory and field results support the hypothesis that NOM migration can be described as a multicomponent transport process. In laboratory studies, hydrophilic subcomponents of NOM (defined by fractionation of NOM using XAD-8 chromatography; Leenheer 1980) had lower sorption affinities for soils and aquifer sediments and were transported more rapidly through soil columns compared with hydrophobic subcomponents of the NOM (Dunnivant et al. 1991). Likewise, in this field study, differences were observed in the mobility of different subcomponents of NOM. Fractionation of the water recovered from the sampling wells demonstrated that hydrophilic NOM was less retarded than hydrophobic fractions. Different size components of NOM also showed differences in mobility. Small NOM molecules (<3000 Dalton mol wt as defined by filtration using Amicon hollow fiber filters) were less retarded than was larger NOM (3 to 100 kiloDalton).

A small amount of the rapidly transported NOM was >0.1μm in size. It is postulated that this NOM is sorbed to, and cotransported with, iron colloids; the iron colloids were approximately 200 nm in size (by photon correlation spectroscopy and scanning electron microscopy) and were negatively charged (by microelectrophoretic mobility).

(5) Desorption of the NOM from the aquifer material is very slow. In laboratory columns studies, there was an immediate and steep drop in NOM eluted from the columns when the input to the column was changed from to a NOM solution to a solution identical in ionic strength and pH but without NOM (Dunnivant et al. 1991). Similarly, in the forced- gradi-

ent field experiment, there was a rapid decrease in NOM at the sampling wells when the NOM injection is terminated at 312 h. NOM does continue to desorb over long periods. At 1248 h, the NOM concentration remains approximately fourfold higher than that of the desorbing groundwater injection solution or the ambient NOM concentration prior to the experiment. A significant amount of the mobile NOM is >3 kiloDalton in size.

In summary, NOM was mobile in a forced-gradient injection experiment. Laboratory studies predicted most of the features in the breakthrough curves, including evidence of a slow kinetic component to DOC sorption, very slow desorption of NOM, and differences in mobility of different subcomponents of NOM (defined by size and sorption to XAD-8).

Changes in NOM and dissolved oxygen levels also affected iron dynamics, but these results are described in Liang et al. 1992 in this symposium proceedings.

REFERENCES

Dunnivant, F.M., Jardine, P.M., Taylor, D.L., and McCarthy, J.F., Transport of naturally-occurring dissolved organic carbon in laboratory columns containing aquifer material, *Soil Sci. Soc. Amer. J.,* (in press), 1991.

Jardine, P.M., Dunnivant, F.M., Selim, H.M., and McCarthy, J.F., Comparison of Models for describing the transport of dissolved organic carbon in laboratory columns containing aquifer material, *Soil Sci. Soc. Amer. J.,* (in press), 1991.

Jardine, P.M., Weber, N.L., and McCarthy, J.F., Mechanisms of dissolved organic carbon adsorption on soil, *Soil Sci. Soc. Am. J.,* 53:1378-85, 1989.

Leenheer, J.A. and Huffman, E.W.D., U.S. Geol. Surv., Water-Resour. Invest., 1979.

Liang, L., McCarthy, J.F., Williams, T.M., and Jolley, L., Iron dynamics during injection of dissolved organic carbon in a sandy aquifer, in *Manipulating Groundwater Colloids for Environmental Restoration,* McCarthy, J.F. and Wobber, F.J., Eds., Lewis Publishers, Chelsea, Michigan, 1992.

CHAPTER 7

Hydrogeochemical Factors Influencing Colloids in Waste Disposal Environments

Philip M. Gschwend

A variety of changes may occur in a subsurface environment in response to waste disposal. These changes may involve physical factors, such as altered flow velocities, or biologically mediated ones, such as the formation of biofilms. Also, new inputs of a variety of chemicals may occur. The combination of these influences often causes the subsurface solution properties to change, and as a result, the formation, mobilization, or stabilization of colloids in groundwater is likely to be affected. To anticipate the magnitude of these effects, it is useful to review how extensively various critical solution properties change in response to waste disposal. This paper briefly summarizes reported changes in groundwater temperature, pH, E_H and O_2, ionic strength and Cl^-, specifically sorbed cations like Ca^{+2}, and total organic carbon (TOC) for a variety of geological and waste-type cases (Table 1) and notes the potential impacts these changes could have on groundwater colloids.

Groundwater temperatures have been seen to increase, decrease, or remain unchanged in response to waste disposal activities. Maximum changes in the near-field have been reported to be as much as several degrees centigrade (Nicholson et al. 1983; Penrose et al. 1990). Probably this would cause only minor changes with regard to factors influencing colloids. For example, cementitious secondary calcite solubility varies only a few percent over such a temperature range; also the rate of hetero-coagulation processes would probably change by less than a factor of two as a result of a 5°C temperature change (assuming an attraction energy barrier of 40 kJ/mol or less).

Table 1. Brief survey of geochemical changes at a variety of waste disposal sites.

Waste Type		Location	Reference
(1) treated sewage infiltration	e.g.,	Cape Cod, MA Palo Alto, CA Ft. Devens, MA	LeBlanc, 1982 Roberts et al., 1982 Satterwhite et al., 1976
(2) landfills	e.g.,	Borden, Canada Toronto, Canada Australia	Nicholson et al., 1983 Barker et al., 1986 Barber et al., 1990
(3) fly ash burial	e.g.,	Nevada	Gschwend et al., 1990
(4) coal tar deposits	e.g.,	New York Connecticut	Groher, 1990 Backhus, 1990
(5) mine tailings, uranium	e.g.,	Ontario, Canada	Morin et al., 1988
(6) rad-waste water	e.g.,	Los Alamos, NM	Penrose et al., 1990

Groundwater pH near waste sites has been seen to increase (LeBlanc 1982; Barker and Barbash 1988; Groher 1990; Backhus 1990; Penrose et al. 1990) as well as decrease (Nicholson et al. 1983; Gschwend et al. 1990; Morin et al. 1988). The extent of change is often about 1 pH unit, but big variations have been reported, such as when waste materials are limed (Penrose et al. 1990) or when sulfide oxidation such as in some mine tailings enables substantial sulfuric acid formation (Morin et al. 1988). Although waste inputs may introduce large quantities of acid or base to the subsurface, waste interactions with the soils and aquifer solids work to resist pH change. There are at least two major influences of pH change on subsurface colloids. First, due to pH changes, the surface charge densities of minerals will vary, and as a result electrostatic interactions of colloids with one another and with immobile aquifer grains will change correspondingly. Thus, coagulation may be promoted or retarded depending on the solids (and their pH_{zpc}) of interest. The second major effect involves the solubilization of subsurface solids. This may be particularly important to soil colloids if cementitious iron oxides or calcium carbonate phases are dissolved at lowered pHs (e.g., Gschwend et al. 1990).

Another common change in groundwater chemistry resulting from waste introduction involves lowering the solution E_H and dissolved oxygen content. This effect has been reported at sewage infiltration sites (LeBlanc 1982; Magaritz et al. 1990), landfills (Nicholson et al. 1983), and coal tar sites (Groher 1990; Backhus 1990), undoubtedly because of inputs of degradable organic matter. A major impact of changed redox conditions on colloids occurs through processes involving subsurface iron. Secondary iron oxide phases may dissolve, thereby mobilizing previously bound colloids (Ryan and Gschwend 1990); and specifically sorbed Fe^{+3} on silicates would be reduced and desorbed, thereby increasing silicate mineral surface charge and electrostatic repulsion. In one instance, production of ferrous iron appears to have been sufficient to enable ferrous phosphate colloids to form (Gschwend and Reynolds 1987).

Still another factor changed by waste disposal is groundwater salt content, usually causing subsurface solutions to exhibit higher ionic strengths. This has been seen at landfills (Barker and Barbash 1988; Barber et al. 1990; Nicholson et al. 1983), at fly ash disposal sites (Gschwend et al. 1990), and at sewage infiltration locations (LeBlanc 1982) and typically involves increasing ionic strengths above 10 mM. There are some instances where wastewater inputs cause groundwater to become less salty, as was the case where seawater infiltration at a coastal site was reversed (Roberts et al. 1982). Obviously, changed groundwater ionic strength affects changed constituent reactivities through activity coefficients. Also, elevated ionic strength causes a shrinkage of the diffuse double layer around particle surfaces, thereby promoting coagulation (e.g., Kia et al. 1987). In cases where groundwater is freshened by recharge, colloid mobilization may be seen to result from the opposite effect (Nightingale and Bianchi 1977).

An important, possibly more subtle, change involves the introduction of specifically sorbed ions like calcium with the waste. In the case of calcium, its concentration has typically been seen to increase with sewage infiltration (LeBlanc 1982) and near some landfills (Nicholson et al. 1983); in instances where solution pH is increased, carbonate precipitation may lower calcium level (Barker and Barbash 1988; Penrose et al. 1990). Because ions such as Ca^{++} are strongly sorbed to minerals like silicates and aluminosilicates, they may act to encourage coagulation at level at which surface charge densities are minimized (Iler 1975; Delgado et al. 1986).

A final major solution effect of note involves the introduction of organic carbon to the subsurface. At landfills, this input can be dramatic (Barker and Barbash 1988; Barber et al. 1990; Nicholson et al. 1983), although much of this organic material may be volatile fatty acids. To the extent that macromolecular organic matter is added, it may coat surfaces of both immobile aquifer solids and any colloids in suspension. This could cause stabilization of suspended colloids (O'Melia 1990) and diminish their tendency to become attached to immobile aquifer grains. Such a stabilization process must be involved to explain the microelectrophoresis behavior of the ferrous phosphate colloids found in a sewage-derived groundwater plume (Gschwend and Reynolds 1987). The association of macromolecular organic matter with colloidal species may also enhance the sorptive capacity of such microparticles for some kinds of contaminants.

In summary, waste discharge to subsurface environments often causes significant changes in solution properties, which may, in turn, dictate the importance of colloid-mediated processes. These effects are manifested by changing solid solubilities, surface charge densities, and sorbent characteristics. Future research needs to look at the condition of solid phases as well as solutions near waste disposal sites.

REFERENCES

Backhus, D.A., Colloids in groundwater: Laboratory and field studies of their influence on hydrophobic organic contaminants, Ph.D. Thesis, Department of Civil Engineering, Massachusetts Institute of Technology, 1990, 198.

Barker, J.F. and Barbash, J.E., *J. Contam. Hydrol.*, 3:1, 1988.

Barber, C., Davis, G.B., Briegel, D., and Ward, J.K., *J. Contam. Hydrol.*, 5:155, 1990.

Delgado, A., Gonsalez-Caballero, F., and Bruque, J.M., *J. Colloid Interface Sci.*, 113:203, 1986.

Groher, D.M., An investigation of factors affecting the concentrations of polycyclic aromatic hydrocarbons in groundwater at coal tar waste sites. M.S. Thesis, Department of Civil Engineering, Massachusetts Institute of Technology, 1990, 145.

Gschwend, P.M., Backhus, D., MacFarlane, J.K., and Page, A.L., *J. Contam. Hydrol.*, 6:307, 1990.

Gschwend, P.M. and Reynolds, M.D., *J. Contam. Hydrol.*, 1:329, 1987.

Iler, R.K., *J. Colloid Interface Sci.*, 53:476, 1975.

Kia, S.F., Fogler, H.S., and Reed, M.G., *J. Colloid Interface Sci.*, 118:58, 1987.

LeBlanc, D.R., Sewage plume in a sand and gravel aquifer, Cape Cod, Massachusetts. Open-File Report 82-274. U.S. Geological Survey, 1982, 35 pp.

Magaritz, M., Amiel, A.J., Ronen, D., and Wellz, M.C., *J. Contam. Hydrol.*, 5:333, 1990.

Morin, K.A., Cherry, J.A., Dave, N.K., Lim, T.P., and Vivyurka, A.J., *J. Contam. Hydrol.*, 2:271, 1988.

Nicholson, R.V., Cherry, J.A., and Reardon, E.J., *J. Hydrol.*, 63:131, 1983.

Nightingale, H.I. and Bianchi, W.C., *Ground Water*, 15:146, 1977.

O'Melia, C.R., Kinetics of colloid chemical processes in aquatic system, in *Aquatic Chemical Kinetics*, Stumm, W., Ed., Wiley-Interscience, New York, 1990, 447-74.

Penrose, W.R., Polzer, W.L., Essington, E.H., Nelson, D.M., and Orlandini, K.A., *Environ. Sci. Technol.*, 24:228, 1990.

Roberts, P.V., Schreiner, J., and Hopkins, G.D., *Water Res.*, 26:1025, 1982.

Ryan, J.N. and Gschwend, P.M., *Water Resources Res.*, 26:307, 1990.

Satterwhite, M., Condike, B., Stewart, G., and Vlach, E., Rapid infiltration of primary sewage effluent at Ft. Devens, MA, CRREL Report No. 76-48. 1976, 34 pp.

SUMMARY AND RESULTS OF

SECTION I

MOBILIZING AND DEPOSITING COLLOIDS

CHAPTER 8

Discussion Leaders

Philip M. Gschwend and John M. Zachara

The group discussion focused on one major remediation strategy: mobilizing groundwater colloids to enhance pump-and-treat approaches to environmental restoration. Within this context, four approaches to mobilizing particles were considered in detail and are outlined in Table 8.1. The following expanded discussion and description integrate consideration of all the processes which are all based on similar theory.

APPROACH: MOBILIZE COLLOIDS FOR PUMP-AND-TREAT

Description

Many contaminants appear to be difficult to flush from groundwater environments because of their tendency to sorb to subsurface solids. If the primary sorbents are subsurface colloids like clays, iron oxides, or humic substances, flushing effectiveness could be enhanced by causing these colloidal phases, and their contaminant sorbates, to move with the flowing groundwater.

Processes

Processes proposed for mobilizing colloids fall into four categories:

- Dissolution of secondary mineral phases "cementing" colloids to aquifer media.
- Increasing the charge development by pH manipulations to promote detachment and mobilization of colloids.

- Application of specific sorbates to alter surface charge on colloids and/or on media.
- Dispersion of colloidal particles by lowering the ionic strength of groundwater.

Many subsurfaces colloids appear to be held in place by secondary minerals such as iron oxides. To release such colloids from the immobile structural grains, the groundwater chemistry could be modified to promote dissolution of such secondary phases. Lowering the solution Eh would drive reduction of Fe^{III} to the much more soluble Fe^{II} thereby dissolving cementitious iron oxides and freeing included colloidal clays or humic substances.

Another common secondary cementitious phase is calcium carbonate; hence, acidification of groundwater would dissolve such solids and free colloids bound up with them. Additionally, changing solution pH has the effect of varying the extent of charge development on all of the solids in the system. By establishing a groundwater pH that dictates maximal like-charge density on colloids of interest and immobile structural grains, one would promote the detachment and subsequent transportability of the mobilized colloids.

Application of specific sorbates such as surfactants to a particle milieu such as the subsurface may result in all solid surfaces acquiring like charge. That is, while quartz may be negatively charged under ambient conditions, adsorption of anionic surfactants to aluminum or iron oxides could convert them from positively- to negatively-charged particles. Subsequent mobilization and transport of colloidal phases would thus depend on the resulting electrostatic repulsions to prevent re-attachment of suspended colloids.

In some cases, colloids such as clays may be immobilized through van der Waals attractions that exceed electrostatic repulsions, reduced by narrow, diffuse double layers. If subsurface ionic strength were lowered with the concomitant expansion of double-layer thicknesses around the clays, these aluminosilicate colloids could be dispersed into the flowing groundwater.

Advantages

Application of chemical reductants may allow good control over solid phases that are likely to react. Some experience with similar subsurface solid phase dissolution has already been acquired in the mining industry. Removal of secondary phases may also increase aquifer permeability and, thus, flushability. The use of organic substrates and a specific terminal electron acceptor to enable in situ microorganisms to poise the Eh may limit adverse side-effects. It seems likely that suitable microorganisms would be present naturally. Dissolution of iron oxides may also promote recovery of coprecipitates or sorbed pollutants (e.g., other metals).

As for chemical reductants/oxidants, application of defined acids or gases to the subsurface may permit a certain level of control on the resultant subsurface conditions. In addition, strategies on the use of surfactants may benefit from the experiences of the cleaning industry.

A major advantage of low-ionic-strength water flushing is that no undesirable chemicals are added that could potentially cause adverse side effects in this regard, some experience in the petroleum industry and the waste-water injection field may assist in the design and feasibility assessment of this approach.

Limitations

It is unclear how quickly changes in groundwater chemistry will translate into colloid releases. This is especially true given the multiphasic nature of subsurface solids, such that changing conditions to mobilize one solid phase may simultaneously affect the behavior if a second solid and thus its roles as a heterocoagulant. Furthermore, it is likely that colloidal phases are immobilized by a variety of mechanisms at different locations. Consequently, no one approach is likely to work everywhere.

Except for flushing with low ionic strength water, manipulations of groundwater chemistry may require release of chemicals that are themselves toxic.

Gaps in Knowledge

It is unclear how quickly subsurface iron oxides would dissolve and release co-coagulated colloids in response to changes in groundwater Eh. A similar comment can be made with regard to dissolving secondary calcium carbonates. Thus, studies of the factors limiting secondary phase reactivity and of the mechanisms by which these phases immobilize colloids are needed. These, in turn, should facilitate predictions of colloid release rates.

Accurately predicting the extent of association of specific sorbates such as surfactants with natural heterogeneous solids is not currently possible. Additional research on the fundamental mechanisms of amphiphile sorption is necessary, especially real-world solids. All of the approaches suffer from our uncertain ability to "adjust" groundwater conditions as we would like in a subsurface. Further, many of these "additive" approaches must be cautiously examined for side-effects because of introduction of residuals after cleanup is complete.

Overall Feasibility

The group did not have enough time to ''rank'' various mobilization approaches by feasibility. The feasibility of any particular approach will be site specific. Further, the potential for undesirable side-effects will all be site specific. Given that pump-and-treat approaches are quite common and that current experiences have not been especially successful, attempt to manipulate colloidal phases must be viewed with caution.

DISCUSSIONS AND CONCLUSIONS

Various of theoretical and empirical (from laboratory experience) insights suggest that engineered mobilization of colloidal phases in subsurface environments may be feasible. Our *field* experience in this vein is quite limited, however. It is clear that a few different mechanisms are probably responsible for immobilizing colloids at different sites: heterocoagulation with secondary mineral phases, heterocoagulation in response to oppositely charged solids attracting, coagulation in conditions of low surface potentials and high ionic strengths. It is also clear that many contaminants of concern are primarily bound to colloidal phases. Thus, efforts to pump-and-treat such contaminants would probably be much improved if we could dissolve or mobilize the relevant colloids.

Several recommendations for future work are

- pursue continued investigation of the mechanism(s) immobilizing colloids in existing subsurface environments;
- clarify the transportability of suspended colloids, especially regarding their ''sticking efficiency'' under various solution conditions;
- study the nature of specific adsorption, especially of surfactants and especially on natural solids, so that their influence on subsurface properties can be anticipated; and
- attempt pilot-scale mobilization of colloids at various sites to learn of the difficulties and unanticipated side effects.

TABLE 8.1 Evaluation of strategies considered in group discussions.

PROPOSED MANIPULATION APPROACH	KEY GEO-HYDRO-BIO-CHEMICAL PROCESSES IN APPROACH	ADVANTAGES/ REASONABLENESS OF APPROACH	LIMITATIONS/ IMPRACTICALITIES IN APPROACH	MAJOR GAPS IN KNOWLEDGE	OVERALL FEASIBILITY
1. Mobilize colloids by dissolution of cementing phases	Dissolution of secondary mineral phases Reductive dissolution of iron oxides Acid dissolution of carbonates	Some experience in mining industry Dissolution may increase permeability and flushability	Kinetics of colloid release uncertain Secondary effects on other solids may be complex Probably site specific	Dissolution kinetics Predictability	Appears promising and worthy of pilot-scale trials
2. pH manipulations to alter surface charge	Surface charge development on colloids and media Electrostatic stabilization	Theoretical framework developed	Chemical heterogeneity (micro- and macro-scale) in natural systems Site specific	Predictability	May be feasible in some sites
3. Specific sorbates to alter surface charge	Adsorption on colloids and media Stability of colloids	Experience in cleaning industry More targeted than pH/Eh manipulations	Undesirable effects of chemicals Microbial degradation of surfactants Possibility of multiple mechanisms	Predictability Mechanisms of amphiphile sorption	
4. Manipulate ionic strength to control dispersion/ coagulation	Dispersion Coagulation	Theoretical framework developed No undesirable chemicals introduced Experience in petroleum industry and wastewater injection field	Flow heterogeneity and mixing at the pore scale Possible permeability reductions due to pore clogging may limit treatment		

SECTION II

BIOCOLLOID MOBILITY

Limits on Quantitative Descriptions of Biocolloid Mobility in Contaminated Groundwater

Ronald W. Harvey and Edward J. Bouwer

MODERATOR'S OVERVIEW

General Considerations in Biocolloid Mobility

Widespread and ongoing problems involving groundwater contamination have led to an increasing interest in factors that affecting biocolloid mobility in the subsurface. It is believed that degradation of highly mobile and persistent groundwater contaminants may be enhanced by cotransport of bacteria that have become acclimated to the presence of these compounds. Also, recent advances in genetic manipulation of bacteria in pure culture have improved the feasibility of using genetically engineered microorganisms (GEMs) for biorestoration of organically-contaminated aquifers. However, the effectiveness of GEMs or waste-adapted bacteria in situ treatment schemes depends, in part, on their ability to reach the contaminant-affected zones of the aquifer.

In contrast, movement of pathogenic bacteria, protozoa, and viruses through the subsurface has long been a topic of research, because contamination of water-supply wells by microbial pathogens contributes substantially to waterborne disease outbreaks. A common problem involving domestic use of untreated groundwater is the subsurface transport of disease-causing bacteria and viruses from septic tanks, land fills, domestic waste lagoons, and on-land disposal facilities for treated and untreated sewage effluent. Although the ability of biocolloids to move through fractured bedrock and limestone has been known for some time, a growing number of reports suggest that pathogen and "indicator" bacteria, under some conditions, can travel for long distances (several hundred to a thousand meters) through a variety of aquifers, including pebbles, gravels, sand and gravel, stony silt loam, and even fine sand. However, a number of these observations are difficult to verify because the source term(s) are not well-defined.

Both conceptual and data-based approaches have been employed in the construction of models for describing transport of biocolloids through the subsurface. Theoretical models for describing microbial transport through porous media were developed several years ago. However, verification using field observation is difficult due, in part, to the model's complexity and the number of parameters that would need to be determined a-priori. Also, some evidence suggests that the predictive value of such models is limited. In general, predictive modeling of microorganisms through contaminated groundwater is problematic because of the number of processes involved, many of which are poorly understood. Because the extent of pathogen movement through the subsurface can be influenced as much by factors controlling survival as those that control their mobility, modeling efforts involving virus transport have emphasized both persistence and transport. More accurate predictive models for virus transport through the subsurface are being constructed and should be available shortly.

Accurate models for describing bacterial transport through the subsurface may not be available for some time, although bacterial transport has been modeled in recent flow-through column experiments and small-scale, in situ tracer tests. Models describing bacterial transport through contaminated aquifers can be affected by a number of biotic and abiotic factors in addition to the geologic and hydrologic characteristics of the aquifer itself. Several factors that govern bacterial processes may be inter-related and difficult to describe mathematically. For example, the chemical conditions in the groundwater may affect several processes (e.g., growth, survival, attachment, and detachment) that, in turn, affect the flux of bacteria through an aquifer.

Knowledge gaps among biological processes currently hampering the quantitative description of bacterial transport in groundwater include: the role of subsurface protozoa; the magnitude of and controls on in situ bacterial growth; the roles of competitive, parasitic, and predatory microorganisms; the relationships between changes in bacterial size and/or cell surface characteristics and changes in groundwater conditions (particularly in nutrient availability); the controls on irreversible and reversible attachment; and the factors that control the survival of indigenous and non-indigenous bacteria, including GEMs. Very little is known about the subsurface protozoan community. Although protozoa are capable of significantly reducing bacterial populations when present in appreciable numbers, their abundance, distribution, and ecology in aquifers are not well understood. Protozoa may be particularly important in the survival of non-indigenous (introduced) species. Bacterial growth in most aquifers appears to be quite slow and the ability to control bacterial growth rates in situ is not well understood. However, in aquifer restoration schemes that depend on bacterial movement to and through zones of contamination, bacterial growth may be required to make up for the losses and immobilization occurring during transport downgradient. Phages and Bdellovibrio are known to be present in the subsurface environment, but their contribution in the mortality of indigenous and introduced bacteria is not understood. However, parasites and grazers are commonly very important in the population dynamics of bacterial populations in surface water environments. A number of bacteria have been shown to alter their surface chemistry and overall size in response to changes in nutrient conditions. Copiotrophic marine bacteria commonly respond to starvation conditions by dwarfing and becoming more hydrophobic. In contrast, some soils bacteria appear to become more hydrophilic in response to low abundance of nutrient. The response of bacteria in the aquifer to changes in availability of nutrients has not been investigated. The survival of GEMs in the subsurface is not known, but is presently being investigated in a number of laboratories.

Knowledge gaps involving a biological processes in modeling bacterial transport in aquifers can include: the relative contributions of straining and sorption in immobilization in heterogeneous porous media characterized by widely-varying grain size, the use of filtration theory in describing bacterial immobilization, the handling of physical heterogeneity in transport models, the sorption process itself, and the role of secondary (preferred flow-path) pore structure.

The modeling of bacterial transport is further complicated by the existence of several modes of bacterial migration. Three non-diffusive mechanisms are continuous (unretarded) transport, intermittent (retarded) transport, and taxis. Although all three modes may be common to both indigenous and introduced bacteria in the aquifer, the contribution of each may be poorly understood and can depend upon the geohydrology, mineralogy, type of bacteria, degree of contamination, and physicochemical conditions. Therefore, the ability to manipulate biocolloids for the purposes of aquifer restoration and better predictive modeling of pathogen mobility in groundwater may require more information relating to such topics as the reversibility of attachment, the population dynamics and microbial ecology of the subsurface, the importance of protozoa and bacterial taxis in groundwater, the effect of nutrient conditions upon cell hydrophobicity and size, and the survival characteristics of bacteria that have been introduced into the aquifer. Substantial laboratory and *in-situ* experimentation will be required in order to understand several of these factors to the point of being able to reasonably describe them mathematically or to integrate them into a transport model.

Consideration of Specific Processes Affecting Biocolloid Mobility

The objective of Session II is to present information on what controls mobility of microorganisms in porous and fractured media, and to discuss whether this mobility can be manipulated to improve delivery of microbes for bioremediation. In the discussion session, several specific processes that affect biocolloid mobility are considered. The remainder of this overview will highlight the key processes involved in the movement of microorganisms through porous media and will identify important knowledge gaps.

Fundamental Processes

Microorganisms and their exudates (e.g., metabolites and exocellular polymeric substances, EPS) are important colloids in the subsurface. The activity of the microorganisms results in biotransformation of compounds, many of which are contaminants. Close association between the contaminants and microorganisms is critical to the success of bioremediation. The exudates may be important sources of organic material that can sequester metals and other organics. The movement of pathogens is a public health issue. The fundamental processes involved in the movement of microorganisms in porous media and eventual accumulation of biofilm are illustrated in Figure 1. The accumulation of subsurface biofilms for the purpose of bioremediation is the net result of *transport, attachment, detachment,* and *growth.* The rates of biofilm accumulation and contaminant biotransformation are strongly influenced by transport characteristics, including pore velocity distribution, dispersivity, molecular diffusion, surface roughness, and other variables that affect the delivery rate of cells, substrate, and nutrients.

Several knowledge gaps exist for the set of complex processes illustrated in Figure 1. First, how are the processes of transport, attachment, detachment, and growth interrelated? For example, the accumulation of biomass in porous media influences the hydraulic conductivity, which alters the transport characteristics. Second, mathematical descriptions of the processes need to be developed so that comprehensive models can be formulated.

The sequence of events that leads to microbial colonization of surfaces is shown in Figure 2. Transport of cells to a surface is the initial step and occurs by three different modes. One mode is diffusion. Here the kinetic

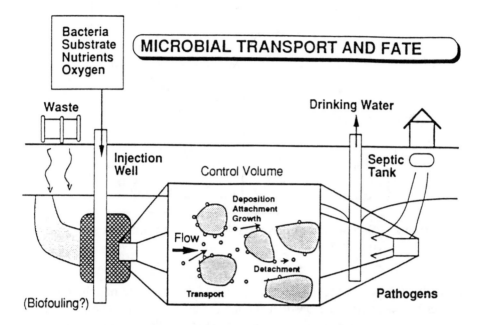

Figure 1. Processes controlling the accumulation of biomass in the sub-surface.

energy of water molecules is transferred to the microorganisms by continu-ous bombardment of water molecules. This motion accounts for random contacts of microorganisms with the surface. The second mode is advec-tion or convective transport. Here fluid flow carries cells in contact with the surface, and the rates may be several orders of magnitude faster than diffusive transport. The third mode is active movement. Some microor-ganisms are motile, and they may chemotactically respond to concentration gradients near a surface. Following transport, the initial association with the surface is mainly a physicochemical process. Theories of colloid and surface chemistry have been successful in describing some microbe-surface interactions. The early stages of microbial attachment can be understood in terms of hydrophobic attraction and electrostatic repulsion. After the microorganism has been deposited on the solid surface, special cell surface structures, such as polymers (EPS) and fibrils, may form strong links be-tween the cells and the solid surface. Finally, microbial growth leads to surface colonization and biofilm formation. Part of the newly formed cells may be released into the bulk liquid.

One gap in our understanding of the events that lead to microbial colo-nization of a surface is the relative contribution of taxis, diffusion, and advection to overall transport. How important is motility in the presence

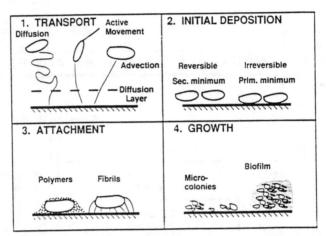

Figure 2.   Sequence of events that lead to colonization of surfaces by microorganisms.

of fluid flow? A second knowledge gap is the influence of solution and surface chemistry on deposition of microorganisms. Chemical aspects play a strong role in the stability of microorganisms. Divalent metal ions tend to decrease particle stability and organic macromolecules tend to stabilize particles. Metabolic activity alters chemical speciation and functional groups on surfaces. Our knowledge of such chemical factors and their relation to microbial deposition and attachment is poor. A related third knowledge gap is the influence of cell surface properties on deposition. Finally, how applicable are models for non-biological colloids to microorganisms? Microbial attachment is generally qualitatively consistent with electrostatic and hydrophobic theories, but how quantitatively accurate are the theories?

Influences of Solid Surfaces on Microbial Activity

There are numerous reports in the literature that attachment to a surface can undoubtedly affect the activity of microorganisms, although sometimes in ways that are not readily predictable with our current knowledge. Examples of effects include increased growth rates, decreased growth rates, decreased substrate utilization (mass transfer resistance), increased substrate utilization (accumulation of substrates at the surface), increased genetic exchange, change in pH optimum, decreased motility, and no influence. In many cases, the data can be rationalized in terms of poorly controlled transport. Because the subsurface is characterized by high surface area, microorganism-surface interactions are critical to biotransformation of contaminants.

For a proper interpretation of possible effects that solid surfaces can exert on microbial metabolism, it is important to consider the geometry of the interaction range between cells and surfaces (Figure 3). Bacteria are 1-2 μm in diameter, and have cell walls of 20-100 nm thickness. The electrostatic interaction between a cell and the surface occurs over a distance of a few nm. Similarly, the thickness of a sorbed layer of substrates and nutrients usually does not exceed 10 nm. This geometry implies that only a very small part of the bacterial surface (<0.1) is in direct contact with the solid surface and sorbed materials. Consequently, the influence of any surface appears to be predominantly indirect. Direct influences of a surface on activity, if any, are expected only for a small fraction of the bacterial surface. Hence, if direct influences of the presence of a nearby surface do exist at all, they are likely outweighed by indirect ones. This provides confidence in applying microbial kinetic data obtained in suspended cultures to the activity of biofilms.

More information is needed to delineate whether surfaces influence microbial activity by direct or indirect means and how these influences affect bioremediation for contaminant control. Biotransformation rates appear to be determined by the solution concentration. Because partitioning of contaminants onto aquifer solids reduces the bulk liquid concentration, the overall rate of reaction is reduced. The availability of sorbed substrates for biotransformation is an important knowledge gap. Consumption of the contaminant decreases the liquid phase concentration, allowing desorption to occur. The overall rate of contaminant utilization is then a function of the desorption rate and microbial kinetics.

Detachment

Detachment is the loss of cells from the surface. The rate of detachment influences the time to accumulate biomass and to achieve a given biotransformation. The mechanisms that contribute to detachment are grazing, erosion, abrasion, and sloughing. Grazing results from predator feeding on the outer surface of microbial colonies or biofilm. Erosion is the continuous removal of biocolloids from the attached biomass, primarily caused by the shear stress created by water flowing past the surface. Abrasion is caused by the collision and or rubbing of biomass coated particles. Sloughing is the periodic loss of large patches of attached biomass.

Detachment is important to subsurface biocolloids for several reason. First, detachment reduces the accumulation of cells and thus influences biomass activity. Thinner biofilms are created that may reduce the contaminant flux into the biofilm, thereby reducing performance. Second, the concentration of suspended solids in the bulk liquid is increased by detach-

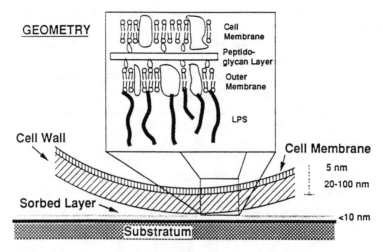

Figure 3.   Geometry of microorganism-surface interaction.

ment. Detached microorganisms could deposit downgradient to expand the zone of biotransformation. Third, as the detachment rate increases, so does the minimum substrate concentration, S_{min}. When S_{min} increases, the lower limit to which a contaminant can be biotransformed increases; thus, it becomes more difficult or even impossible to reduce the contaminant to a given clean-up level. Fourth, detachment losses can affect the species distribution in a multi-species biofilm. The outer layer of microorganisms may be preferentially lost, allowing a sub-layer of microorganisms to proliferate. Finally, detachment can change the surface chemistry of the substratum. Parts of the substratum that were once coated with biofilm can be exposed. This exposure can alter the physical/chemical characteristics of the surface, including diameter, density, surface texture, surface charge, and surface hydrophobicity.

The following are important research questions for detachment processes. What are the local physical, chemical, and biological influences on rates of particle detachment? What mechanisms determine biofilm patchiness in the subsurface? Can sloughing be induced by osmotic pressure gradients and chemical changes? Although some quantitative models exist to describe detachment, better quantitative descriptions are needed. Some success has been obtained in correlating detachment to the growth rate and to the magnitude of hydraulic shear.

It is the objective of this session to review the current state of science with respect to these issues, and to consider in group discussions the key processes which we must understand to effectively manipulate biocolloid mobility for improved environmental restoration.

CHAPTER 10

Movement of Microorganisms Through Unconsolidated Porous Media Without Fluid Flow

Michael J. McInerney, P. Sharma, and P.J. Reynolds

The common assumption in modeling bacterial transport in the subsurface is that the bacterial cells act as particles and are carried through the formation with groundwater flow. However, many bacteria have the ability to move on their own. Taking the fastest velocity recorded for a bacterium (about 50 μm/sec) and assuming that the bacterium moves in a straight line, this would correspond to a velocity of several meters per day, which is equivalent to the average velocity of many groundwaters (1 m per day). The type of analysis indicates that self-propulsion by bacteria may be an important mechanism of penetration in some situations, especially where groundwater velocities are low.

The movement of coliform bacteria often follows the movement of nutrients from leaking septic tanks and sewer lines, suggesting that conditions favorable for bacterial growth are also the conditions that favor greater bacterial penetration. We have shown that bacteria penetrate consolidated and unconsolidated porous media in the absence of fluid flow under conditions that support growth. In one case, the rate of penetration of a motile *Bacillus* species useful in enhanced oil recovery penetrated through Berea sandstone cores with permeabilities greater than 0.1 μm^2 at a rate close to that of most waterfloods (about 0.2 m per day). Interestingly, even a nonmotile bacterium penetrated consolidated sandstone cores. These studies lead us to determine the mechanism(s) and factor(s) important for bacterial penetration in the absence of fluid flow.

The biological factors important for bacterial penetration through anaerobic, nutrient-saturated, sand-packed cores under static conditions were studied using mutants of *Escherichia coli*. Paired comparisons were made between mutant and parental stains to determine how the loss of one function affected penetration. Motile strains of *E. coli* penetrated the cores four times faster than mutants defective only in flagellar synthesis. A nonmotile, gas-producing strain penetrated the cores five to six times faster than did the nonmotile, non-gas-producing mutant, indicating that gas production was an important mechanism for the movement of nonmotile bacteria. The penetration rates of motile, nonchemotactic mutants were 0.16 to 0.28 cm/h, which were four times faster than that of chemotactic parental strains, indicating that chemotaxis may not be required for penetration. For motile strains, strains with the faster growth rates also had faster penetration times, suggesting that the penetration rate is regulated by the in situ bacterial growth rate. A sigmoidal relationship was observed between the specific growth rates of all the motile bacteria used in this study, and the penetration rates through cores saturated with galactose-peptone medium.

To determine why the nonchemotactic strains had faster penetration rates compared with the parental chemotactic strains, cores were sectioned at different time intervals, and the viable number of bacteria in each section was determined. The wild-type, chemotactic strains penetrated the cores in a band-like manner, with high numbers of cells appearing in each new section of the core and without cells being detected in the most distal sections of the core. The nonchemotactic strain, RP 5232, penetrated cores in a more random or diffuse manner. Low numbers of cells were detected in several sections shortly after inoculation and, within 24 h, each section of the core had low but detectable numbers of bacteria.

Measurement of the in situ growth rate and concentrations of substrates and fermentation products showed that the rate of growth in the sand-packed core was slower than in liquid culture and that cell propagation through the core preceded the depletion of the substrate and the large accumulation of products. These studies suggest that self-propulsion by bacteria may be an important mechanism for the movement of bacteria in aquifers with slow flow rates and may provide a mechanism whereby bacteria can propagate throughout the entire porous matrix rather than being localized in the predominant pores channels.

CHAPTER 11

Cells and Surfaces in Groundwater Systems

R.E. Martin, E.J. Bouwer, and L. Hanna

The delivery of microorganisms to contaminated zones in groundwater systems is controlled by hydraulic factors, geological structure, and interactions between the surfaces of suspended cells and the rock or soil. The latter determine the "stickiness" of cells, or their tendency to attach to surfaces. This paper will present a quantitative approach to describing stickiness. The results of experimental investigations using this approach will be discussed. Finally, the potential influence of selected bacterial cell surface characteristics on attachment will be considered.

Although numerous studies of bacterial attachment have given some indication of factors relevant to attachment, they have generally failed to provide a useful parameter for expressing the probability that a cell will attach to a surface with which it collides. This quantity is often referred to as the collision efficiency, or simply α, which is defined as the ratio of the rate at which cells attach to a surface to the rate at which cells are transported to the surface. Because the collision efficiency represents a normalization of the attachment rate with respect to transport, it is independent of the flow conditions and can be used to anticipate deposition behavior in any system that has similar chemical properties.

Purely empirical, semiempirical and purely theoretical approaches to estimating α can be distinguished. The first case entails experimental determination of both the transport and attachment rates; the second consists of measurement of the rate of attachment and calculation of the rate of transport; the third involves calculation of both on the basis of fundamental properties of the system. Currently, theories of particle-surface interactions in aqueous media are not sufficiently developed to make the latter a realistic option. The availability of particle transport models for selected systems, however, makes the semiempirical approach an attractive possibility because it entails half the experimental effort of the fully empirical alternative. It does, however, necessitate validation of the particle transport models.

The choice of experimental systems that lend themselves to theoretical calculation of particle transport as well as measurement of attachment is rather restricted, parallel plate channels and rotating disks being two examples. Experiments designed to obtain collision efficiencies for *P. aeruginosa* depositing on glass surfaces under different electrolyte conditions in a rotating disk system will be described here. The results were used in subsequent experiments to test the use of filtration theory for describing removal of suspended bacteria in a model porous medium.

The rotating disk was selected in part because the transport of cells to the surface can be readily controlled and experimentally measured. It is also very sensitive; the lower sensitivity limit for measurement of deposition by direct counting depends only on the total counted area and the amount of extraneous deposition associated with manipulations of the disk and particle suspensions. A third advantage of the system is that fluid flows toward the disk such that the flux of particles to the surface is relatively high. Experiments with bacteria can be of short duration and still yield significant numbers of deposited cells. Complications associated with growth and die-away over longer periods can thus be avoided. The final major advantage is that the flow in this system is well defined and theoretical models exist for the deposition of Brownian and non-Brownian particles.

The procedure adopted here to measure cell transport to the surface was analogous to that used by others to test particle transport theories. Bacteria are negatively charged, and their accumulation on a positively charged surface in a given time should reflect their rate of transport to the surface. Therefore, glass surfaces were coated with a water-insoluble polymer, 2-vinylpyridine-styrene (2-VPS), that is positively charged in water. In addition to being used for collision-efficiency determinations, estimates of transport rates were used to evaluate particle transport models.

Collision efficiencies for concentrations of sodium chloride from 10^{-6} to $10^{-1}M$ ranged from 8.3×10^{-4} to 4.1×10^{-1}, a trend qualitatively consistent with the theory of double-layer interaction (Figure 1). Abbott et al. (1983) obtained similar results for the deposition of *Streptococcus mutans* in a rotating-disk system.

To control for possible effects resulting from fluid shear, which increases linearly with distance from the center of the disk, counts were made in concentric rings and analyzed as a function of radial position. In Figure 2 the radial distribution of deposition rates, normalized with respect to suspended cell concentration and exposure time for *P. aeruginosa* depositing on 2-VPS at 10^{-6} and $10^{-4}M$ NaCl are shown. The normalized deposition rate decreased by an order of magnitude as the radial position increased from 1 to 10 mm. Similar experiments using polystyrene latex

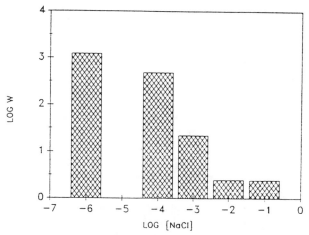

Figure 1. Effect of NaCl concentration on the negative logarithm of the collision efficiency for *P. aeruginosa* depositing on glass in a rotating disk system.

spheres also revealed a radial dependence. The horizontal lines in this plot represent predictions of the trajectory model of Spielman and Fitzpatrick (1973) for 0.7- and 1.6-µm spheres. This model, like all others, predicts uniform deposition. Because bacteria and latex particles behaved quite similarly with respect to the radial variation in deposition, the effect cannot be attributed to particular mechanisms of bacterial attachment. The most likely explanation is nonuniform transport of particles arising from lateral migration or translation of particles across streamlines in the shear flow near the disk surface.

From a practical standpoint, the rotating-disk system has provided a fully empirical approach to determining particle collision efficiencies, provided that deposition is evaluated as a function of distance from the disk center. Because of the radial dependence of particle deposition, the semiempirical approach to determining collision efficiency is not yet feasible. Incorporation of lateral migration into particle deposition models for the rotating disk should be considered to explain the nonuniform deposition. Models for other systems such as parallel plate channels and impinging jets should also be reexamined to determine the importance of migration.

The possibilities for manipulating cell properties to optimize the deposition behavior in a given application consist of altering physical properties, such as size, shape, specific gravity and appendages (fimbriae, flagella), and modification of the chemical properties of cell surface components, including exopolysaccharides.

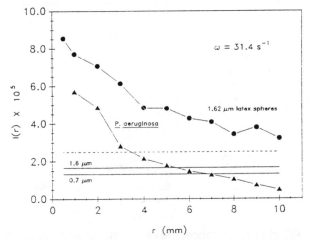

Figure 2. Comparison of radial distributions of normalized deposition rates for 1.62 μm sulfated polystyrene latex spheres and *P. aeruginosa*, 0.7 × 1.6 μm rod shaped bacteria, both deposited on 2-VPS surfaces in a rotating disk system at a rotational speed of 31.4 s⁻¹. The dashed line shows the trajectory model prediction for the latex particles, assuming specific gravity 1.05. The solid lines show the predicted values for 1.6 and 0.7 μm spheres of specific gravity 1.1.

Size, shape and specific gravity would be expected to mainly affect the transport of cells to surfaces and, hence, the collision rate, or the opportunities for cells to be removed from suspension. The impact of changes in these properties on transport, however, depends on the specific hydraulic and geometric properties of the system under consideration; thus it is difficult to generalize about the possibilities for controlling deposition in this way.

With regard to manipulating the collision efficiency, other investigations have shown several factors to be potentially important, especially hydrophobicity and the presence of fimbriae. Very limited published data on the effects of exopolysaccharides (EPS) on attachment suggest relatively little impact on initial attachment, but the production of EPS by attached cells can significantly influence the retention of daughter cells on the surface. In general, however, these studies have been performed under solution conditions bearing little similarity to groundwater and in systems with poorly defined hydrodynamics. Overall, the results suggest that the collision efficiency could vary by as much as a factor of ten with differences in cell surface properties.

The importance of electrostatic forces in the attachment of bacteria to surfaces is clear. Further investigation should be directed toward (1) better understanding of hydrodynamic effects in particle deposition, (2) interactions between these effects and microbial properties, and (3) quantitative assessment of the effects of microbial properties under conditions of solution and surface chemistry relevant to groundwater systems. Finally, high priority should be placed on the development and use of quantitative approaches to assessment of cell stickiness.

REFERENCES

Abbott, A., Rutter, P.R., and Berkeley, R.C.W., The influence of ionic strength, pH and a protein layer on the interaction between *Streptococcus mutans* and glass surfaces. *J. Gen. Microbiol.*, 129:439-45, 1983.

Spielman, L.A. and Fitzpatrick, J.A., Theory for particle collection under London and gravity forces. *J. Colloid Interface Sci.*, 42:607-23, 1973.

CHAPTER 12

Adhesion and Deposition of Bacteria

Huub H.M. Rijnaarts, Willem Norde,
Johannes Lyklema, and Alexander J.B. Zehnder

Adhesion in batch systems has already been shown to be controlled by electrostatic and van der Waals interactions as described by the DLVO theory of colloid stability. The adhesion behavior of microorganisms in column systems is relatively unknown. The objective of this research is to relate the adhesion as determined in column systems to the adhesion as measured in batch system.

Deposition of bacteria in porous media can be regarded as a two-step process. The first step is transport from bulk fluid to collector surface, which is controlled by convection, diffusion van der Waals and electrostatic interactions, the gravitational field, and drag forces. After being transported to the collector surface, adhesion can occur. For both transport and adhesion, an efficiency can be defined. The product of these two factors equals the total deposition efficiency. The efficiency of transport equals the number of particles transported to the surface divided by the total number of microorganisms approaching the collector. The adhesion efficiency equals the number of microorganisms that attach to the surface divided by the number of particles transported to the surface.

Our research strategy consisted of the following: (1) Experiments to determine the total deposition efficiency for column systems utilizing columns packed with spherical collectors. Bacterial suspensions were fed to these columns, and the influent and effluent concentrations were measured. After a number of pore volumes have run through the column, the normalized concentration (which is the effluent over the influent concentration) reaches a stabile value. This stabile concentration decreases exponentially with column length with a decay constant that is called the filtration coefficient. The total deposition efficiency can be calculated from this filtration coefficient. (2) Data from the column studies and equations based on filtration theory were used to calculate transport efficiency. (3) The adhesion efficiency for column systems was calculated from the experimentally determined deposition efficiency and the calculated transport efficiency. (4) The adhesion in batch systems was determined experimentally. (5) The adhesion efficiency in columns was compared with the adhesion in batch systems.

Two Coryneform bacteria species and two Pseudomonads were used. These microorganisms varied in physical/chemical properties as characterized by contact angle and electrophoretic mobility measurements. The surface chemical properties of the substratum were also varied by using glass and teflon collectors and surfaces. The bacteria varied in radius between 0.59 and 1.17 μm. The collector radius was 190 μm and 225 μm for teflon and glass, respectively. For the column experiments an externally imposed flow rate of 27 cm/h and an ionic strength of 10^{-1} M was applied, unless stated otherwise.

As expected, the adhesion efficiency in columns and the adhesion in batch were lower for glass than for teflon. Except for one case, *Pseudomonas putida* mt2 on glass, the adhesion efficiency in columns as function of adhesion as measured in batch seems to follow a curve that levels off at a value of 0.8, when adhesion in batch exceeds 2×10^6 cells/cm^2.

For *Pseudomonas putida* mt2 on teflon, both the adhesion efficiency in columns and the adhesion in batch decrease with decreasing ionic strength.

For columns, the adhesion efficiency increases linearly with increasing filtration coefficient at filtration coefficients below 10 m^{-1}. Above this value the adhesion efficiency levels of at a value of 0.8.

CONCLUSIONS

1. Adhesion efficiency in columns corresponds to the adhesion in batch systems.

2. For predicting the deposition behavior of a bacterial species in porous media, information about the surface chemical properties of the microorganism, collector material, and fluid medium is a prerequisite.

CHAPTER 13

Physical and Chemical Controls on the Advective Transport of Bacteria

A.L. Mills, D.E. Fontes,
G.M. Hornberger, and J.S. Herman

Although models have been developed that describe the transport of bacteria through porous media (Corapcioglu and Haridas 1984, 1985; Peterson and Ward 1989), experimental data are scarce. The models are very difficult to test empirically because of their reliance on a large number of parameters describing geological and bacterial variability, the in situ determinations of which are impractical at best. It is therefore necessary to test first bacterial transport processes under carefully controlled laboratory conditions. The data that currently exist were gathered by using several different methods under dissimilar conditions and a number of different kinds of bacteria. Integration of these disparate data sets is, therefore, not likely to result in the creation of a legitimate model of bacterial transport. Studies involving heterogeneities within the porous media are especially rare, and because the heterogeneities involved were either natural, (e.g., intact soil cores, Smith et al. 1985) or randomly generated heterogeneities (e.g., burrowing earthworms, Madsen and Alexander 1982), they were not characterized. Other studies utilize columns packed with soils of nonuniform size (Wollum and Cassel 1978), confounding analysis of the effects of grain size/pore size domains of the porous media. We have systematically examined the effect of grain size, cell size, and ionic strength in a factorial design to determine the relative importance of each of the variable conditions explored and, eventually, to develop an empirically based, thoroughly tested, reliable model of bacterial transport through porous media.

Bacteria collected from bulk sediments of a freshly hand-augered well were isolated by plating on half-strength peptone/yeast extract (1/2 PYE)

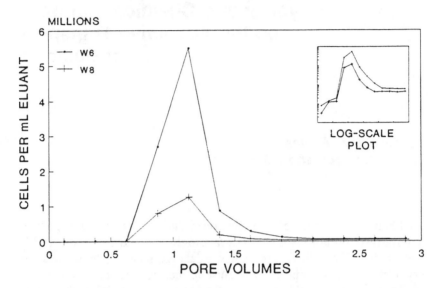

Figure 1.   Breakthrough curve for strains W6 and W8 in the fine homogeneous case with a low-ionic strength eluant. These curves
are typical in shape for all homogeneous case treatments.
The log-scale plot is included to show the magnitude of the
"tail" of bacteria that continue after the pulse of bacteria has
passed. The flat tops observed on some peaks are an artifact
caused by the size of the samples.

agar. Strains W6 and W8 (both isolated from the aqueous portion of the
sample) were selected for use in these experiments on the basis of their
similarities in cell-surface hydrophobicity and gram reaction and their
differences in cell size and shape. Strain W6 is a non-spore-forming gram-
negative coccus (~0.75 μm diam). Strain W8 is a non-spore-forming gram-
negative rod (~0.75 μm × 2.0 μm). Both organisms are highly hydrophilic
(contact angle = 20°) according to the classification scheme of Mozes et
al. (1987).

Resting cells [grown in 1/2 PYE broth, washed free of medium, and
allowed to sit in artificial groundwater (AGW) for 36 h] of the two cultures were passed as a pulse through columns containing acid-washed,
rounded quartz sand. Fine sand was 0.33 mm < diam < 0.40 mm, whereas
coarse sand was 1.00 mm < diam < 1.14 mm. For those experiments
involving structured heterogeneities, a preferred flow path was created by

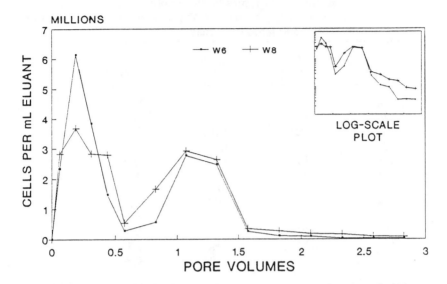

Figure 2. Breakthrough curve for strains W6 and W8 in the heterogeneous with the high ionic strength eluant. Note the double peak in the curve, typical of all experiments with a structured heterogeneity.

inserting a glass tube (0.6-cm ID) in the center of the columns and packing fine sand around it. The inside of the glass tube was then filled with coarse sand, and the tube was carefully removed. Sterile AGW was used as the eluent for the columns. The columns were operated at a flow rate of about 1 pore volume (i.e., 88mL) per hour. Eluent samples were collected from the base of the column in 1/4 or 1/8 pore volume intervals, and analyzed for bacterial concentration using acridine orange direct counts (Hobbie et al. 1977). Experimental runs were continued until at least three pore volumes of AGW had passed through the columns.

Breakthrough curves generated from homogeneous columns were all similar in shape (see Figure 1). In all cases, a "tailing" effect was seen. After reaching a peak concentration at or near one pore volume, bacterial cell number declined and leveled off at a nonzero baseline concentration. This tail (not observed in breakthrough curves for Cl^-) has been shown to persist for at least seven pore volumes (Scholl et al. unpublished data). Breakthrough curves generated from columns containing a preferred flow

Table 1.   Percentage recovery of bacterial cells from columns of clean quartz sand. LIS and HIS refer to low ionic strength (0.001) and high ionic strength (0.01), respectively. CHO and FHO refer to homogeneous columns containing coarse- and fine-grain sand, respectively; HET refers to those columns constructed with a preferred flowpath. W6 and W8 refer to treatments with the small (W6) and large (W8) cells.

STRAIN	IONIC STRENGTH	GRAIN SIZE	PERCENTAGE RECOVERY
W6	LIS	CHO	88.35
W6	LIS	HET	79.15
W6	HIS	CHO	49.30
W8	LIS	CHO	43.55
W8	LIS	HET	39.05
W6	HIS	HET	19.70
W6	LIS	FHO	14.50
W8	HIS	HET	11.70
W8	HIS	CHO	4.45
W8	LIS	FHO	3.90
W6	HIS	FHO	2.75
W8	HIS	FHO	0.34

path exhibited a doubly-peaked pattern (Figure 2). The first peak in concentration occurred well in advance of 1 pore volume (between 0.188 and 0.312 pore volumes). The second peak occurred at around 1.5 pore volumes.

The percentage of cells recovered from the column ranged from less than 1% in the case of strain W8 in fine-grained homogeneous columns at high ionic strength to near 90% for strain W6 in coarse-grained homogeneous columns at low ionic strength.

Increasing the ionic strength of the artificial groundwater (0.001 to 0.01) resulted in a decrease in the percentage recovery of the bacteria. The recovery of strain W6 in coarse homogeneous columns was cut nearly in half; the recovery of strain W8 in fine homogeneous columns was reduced by an order of magnitude by the increase in ionic strength.

In all treatments, the smaller cells (strain W6) exhibited a larger percentage recovery than did the larger cells (strain W8). In all treatments, the percentage recovery of cells from coarse-homogeneous columns was greater than that from fine-homogeneous columns.

Significant differences in percentage recovery were indicated for all treatments (MANOVA, all treatment differences significant at $p > 0.99$). Insight into the relative importance to transport efficiency of each of the variables examined can be gained by ranking all of the treatments in order

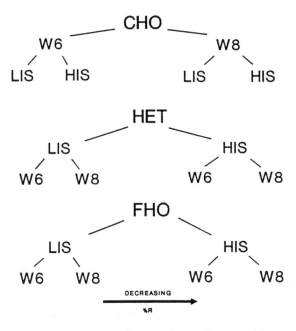

Figure 3.   Pattern of ranking for the percentage recovery of bacterial cells applied to clean quartz sand columns under the various treatment conditions.

of percentage recovery (Table 1). Experimental runs involving low ionic strength, coarse sand, and/or small cells tend to be grouped toward the top of the listing (higher recovery), while those involving high ionic strength, fine sand, and/or large cells are grouped lower on the list (lower recovery). The clearest division appears to be between fine and coarse sands, leading to the conclusion that, among the variables tested, grain size of the porous media may be the dominant factor in determining transport efficiency. By breaking up the listing into coarse-grain columns, heterogeneous columns, and fine-grain columns (Figure 3), we can visually determine what might be the next most important factor. In coarse-sand experiments, a marked division exists between runs involving different cell sizes, while in the fine-sand experiments (as well as those with structured heterogeneities), differing ionic strength appears to be more important than cell size in determining recovery. A response surface analysis including these and other data will allow not only determination of the strength of the effects (Fontes et al. 1991), but will also provide parameter estimates for empirical models calibrated for the magnitude of the treatment variables examined in the work. Detailed analysis of individual treatment variables will provide a refined parameter estimate for such models.

ACKNOWLEDGMENT

This work was supported by Grant No. DE-FG05-89ER60842 from the Subsurface Science Program of the Office of Environmental and Health Research, U.S. Department of Energy.

REFERENCES

Corapcioglu, M.Y. and Haridas, A., Transport and fate of microorganisms in porous media: a theoretical investigation, *J. Hydrol.*, 72:149-69, 1984.

Corapcioglu, M.Y. and Haridas, A., Microbial transport in soils and groundwater: a numerical model, *Adv. Water Resour.*, 8:188-200, 1985.

Fontes, D.E., Mills, A.L., Hornberger, G.M., and Herman, J.S., Physical and chemical factors influencing transport of microorganisms through porous media, *Appl. Environ. Microbiol.*, 57:2473-81, 1991.

Hobbie, J.E, Daley, R.J., and Jasper, S., Use of nuclepore filters for counting bacteria by fluorescence microscopy, *Appl. Environ. Microbiol.*, 33:1225-28, 1977.

Madsen, E.L. and Alexander, M., Transport of *Rhizobium* and *Pseudomonas* through soil, *Soil. Sci. Am. J.*, 46:557-60, 1982.

Mozes, N., Marchal, F., Hermesse, M.P., Van Haecht, J.L., Reuliaux, L., Leonard, A.J., and Rouxhet, P.G., Immobilization of microorganisms by adhesion: interplay of electrostatic and nonelectrostatic interactions, *Biotechnol. Bioeng.*, 30:439-50, 1987.

Peterson, T.C. and Ward, R.C., Development of a bacterial transport model for coarse soils, *Water Resour. Bull.*, 25:349-57, 1989.

Smith, M.S., Thomas, G.W., White, R.E., and Ritonga, D., Transport of *Escherichia coli* through intact and disturbed soil columns, *J. Environ. Qual.*, 14:87-91, 1985.

Wollum II, A.G. and Cassel, D.K., Transport of microorganisms in sand columns, *Soil Sci. Soc. Am. J.*, 42:72-76, 1978.

CHAPTER 14

Modeling Biocolloid Transport in Porous Media: Chemical Aspects and Reversibility

Roger Bales

Attenuation and retardation of biocolloid transport in porous media result from attachment of the colloids to fixed surfaces and subsequent detachment under steady-state conditions and in response to chemical perturbations. Attachment and detachment can be described as first-order processes, with rate coefficients related to microscale physical and chemical properties of the colloid and collector. Governing equations for one-dimensional transport in a porous media with two types of attachment sites, one of which is kinetically limited, are

$$\theta \frac{\partial C}{\partial t} + \phi \frac{\partial S_1}{\partial t} + \phi \frac{\partial S_2}{\partial t} = \theta D \frac{\partial^2 C}{\partial z^2} - u\theta \frac{\partial C}{\partial z} \qquad (1)$$

$$S_1 = fK_p C \qquad (2)$$

$$\phi \frac{\partial S_2}{\partial t} = \theta k_1 C - \phi k_2 S_2 \qquad (3)$$

where C is the colloid concentration in the aqueous phase; S_1 and S_2 are the bound concentrations for fast and kinetically limited sites, respectively;

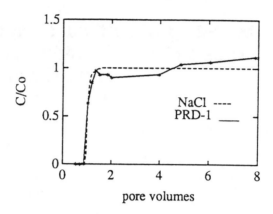

Figure 1. Breakthrough curve for PRD-1 bacteriophage at pH 7. $u = 4.9$ m dy^{-1}, $C_0 = 1.1 \times 10^5$ pfu mL^{-1}, T = 4°C. $D = 3.1 \times 10^{-5}$ cm^2 s^{-1}. [*Bales et al., 1991*]

θ is porosity; P is the dry bulk density of the solid material; D is the longitudinal dispersion coefficient; u is the average interstitial velocity; k_1 is a pseudo-first-order rate coefficient (s^{-1}) for attachment, which depends on the colloid's molecular diffusion coefficient and the sticking efficiency (i.e., net energy of interaction between colloid and collector); and k_2 is a pseudo-first-order detachment rate coefficient, which also depends on the energy of colloid-collector interaction. If only a small fraction of the surface is covered by attached colloids, the rate coefficients will not depend on the surface-site concentration. Equation (1) expresses the total change in concentration with time resulting from advection, dispersion, attachment, and detachment. Equation (2) expresses the linear attachment/detachment equilibrium for the fast (Type 1, equilibrium) attachment sites. fK_p is the equilibrium partition coefficient for the Type 1 sites, and K_p is the total equilibrium partition coefficient, which would describe the partitioning at $t \to \infty$. Type 1 sites could correspond to colloids held near the surface in a secondary minimum of the potential energy of interaction, with little or no energy barrier for detachment. In Eq. (3), the change in bacteriophage concentration bound to type 2 sites with time is the difference between the attachment and detachment rates.

In the absence of colloid attachment to the solid media, only advection and dispersion affect colloid movement and the equation reduces to

$$\theta\frac{\partial C}{\partial t} + \theta D\frac{\partial^2 C}{\partial z^2} - u\theta\frac{\partial C}{\partial z} \qquad (4)$$

In Figure 1, bacteriophage dispersion is seen to be very similar to that for a salt tracer. Dispersion was estimated using a non-linear-least-squares algorithm (van Genuchten 1981). The advection dispersion equation [Eq. (4)] adequately describes both curves. With reversible attachment of phage to the solid media, the combination of Eq. (1) and (2) gives

$$R\frac{\partial C}{\partial t} + D\frac{\partial^2 C}{\partial z^2} - u\frac{\partial C}{\partial z} \qquad (5)$$

where there retardation factor, $R = 1 + \frac{\phi}{\theta}$. However, the significantly greater apparent dispersion observed in an experiment with phage that attach to the solid media cannot be attributed to mobilephase processes but is, instead, the result of slow adsorption (Figure 2). Although the two-parameter equilibrium model of Eq. (5) gave a good fit, the estimated dispersion coefficient was about 1000-fold too high ($D = 0.089$ vs 3.1×10^{-5} cm^{2}s^{-1} from Figure 1). Describing phage attachment and detachment, as pseudo-first-order processes, with no fast (equilibrium) attachment/detachment, gives an adequate fit to the data. For the first-order model, D is fixed at a value estimated from experiments with no retardation and from salt-tracer data (Figure 1). The parameters R and k_1 are then estimated from the shape and location of the breakthrough curve. For the first-order model, Eqs (1) through (3) reduce to

$$\theta\frac{\partial C}{\partial t} + \phi\frac{\partial S}{\partial t} + \theta D\frac{\partial^2 C}{\partial z^2} - u\theta\frac{\partial C}{\partial z} \qquad (6)$$

$$\phi\frac{\partial S}{\partial t} = \theta k_1 C - \phi k_2 S \qquad (7)$$

The two-site model can also be applied to the data in Figure 2, but in this case the additional parameter failed to provide a significantly better fit. Thus, little can be inferred from the additional parameter.

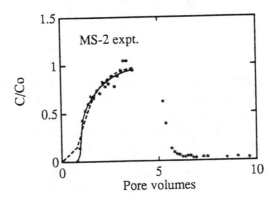

Figure 2. Breakthrough curve for MS-2 bacteriophage at pH 5. $u = 3.2$ m dy^{-1}, $C_0 = 3.5 \times 10^4$ pfu mL^{-1}, T = 4°C. Equilibrium-model fit: D = 0.089 cm^2 s^{-1}, $R = 1.6$. First-order-model fit: $D = 6.9 \times 10^{-5}$ cm^2 s^{-1} (fixed), $R = 1.6$, $k_1 = 2.2 \times 10^{-4}$ s^{-1} [*Bales et al., 1991*]

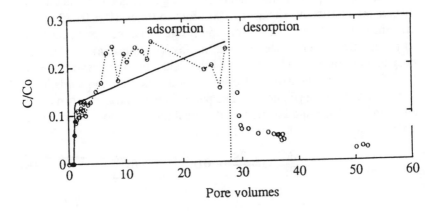

Figure 3. Breakthrough curve for PRD-1 bacteriophage at pH 5.5. $u = 3.2$ m dy^{-1}, $C_0 = 3.5 \times 10^4$ pfu mL^{-1}, T = 4°C. Two-site-model fit: D = 3.1×10^{-5} cm^2 s^{-1} (fixed), $R = 230$, $k_1 = 1.2 \times 10^{-3}$ s^{-1}, $k_2 = 5.6$ $\times 10^{-6}$ s^{-1}. [*Bales et al., 1991*]

Curves that exhibit both some early breakthrough of low phage con-centrations but also exhibit significant apparent phage attachment can best be fit with the four-parameter model of Eqs. (1) through (3) (e.g., Figure 3). The additional parameter f is related to the fraction of equilibrium (fast) vs nonequilibrium (slow) sites on the solid media for colloid attachment/detachment.

In both the first-order and two-site models, the pseudo-first-order rate coefficient k_1 can be interpreted in terms of the rate of colloid transport to a collector times the fraction of colloid-collector collisions that result in attachment (e.g., O'Melia, 1980):

$$\eta \alpha = k_1 C \left(\frac{\pi \theta d^3}{6(1-\theta)} \right) \tag{8}$$

Interpreting the k_1s from the experiments in Figures 2 and 3 in terms of this model, with

$$\eta = [\frac{\mu d_p du}{kT}]^{-2/3} \tag{9}$$

for Brownian transport, gives $\alpha = 4.6 \times 10^{-3}$. Similar α values were observed in other short-pulse experiments (Bales unpublished data) and in longer-term experiments in which the column-outlet concentration reached a steady-state value (Bales unpublished data).

k_1 and k_2, both pseudo-first-order rate coefficients, can be interpreted as time scales for attachment and detachment, respectively. For the MS-2 experiment (Figure 2), $k_1^{-1} = 1.3$ h and $k_2^{-1} = 0.8$ hr for the PRD-1 experiment (Figure 3), $k_1 = 0.2$ h and $k_2^{-1} = 49$ h analysis of replicate curves, including the desorption limbs gave time scales on the order of 3 h for attachment and 30 to 300 h for detachment (Bales unpublished data).

REFERENCES

Bales, R.C., Hinkle, S.R., Kroeger, T.W., Stocking, K., and Gerba, C.P., Bacteriophage adsorption during transport through porous media: Chemical aspects and reversibility, *Environ. Sci. Technol*, (in press), 1991.

O'Melia, C.R., Aquasols: The behavior of small particles in aquatic systems, *Environ. Sci. Technol.*, 14:1052-60, 1980.

Van Genuchten, M.Th., Non-equilibrium transport parameters from miscible displacement experiments, Research Report 119, U.S. Salinity Lab, Riverside, CA, 1981.

SUMMARY AND RESULTS OF

SECTION II

BIOCOLLOID MOBILITY

CHAPTER 15

Discussion Leaders

Edward J. Bouwer and Ronald W. Harvey

The group identified the following four major manipulation approaches that involve microorganisms:

1. Maximize the mobility of microorganisms to inoculate the subsurface.
2. Enhance the accumulation of attached biomass to increase biotransformation rates and to reduce hydraulic conductivity.
3. Promote detachment and removal of accumulated attached biomass to restore permeability and subsurface environmental conditions.
4. Minimize the mobility of pathogens.

Table 15.1 summarizes the consensus of the participants on the key processes and feasibility of these specific strategies. However, the group decided to structure their discussions in a more integrated format that considered the broader issue of **development of biotransformation processes in the subsurface.** The destruction of contaminants during *in-situ* bioremediation is an example of such a biotransformation process. An effective strategy for bioremediation encompasses the key processes and issues identified in the table. The relevance of this topic to colloids in groundwater is several-fold:

- First, microorganisms fall within the colloidal size range and are important colloids in groundwater that can be transported with bulk liquid flow to reach and biodegrade contaminants.
- Second, biomass accumulated on surfaces (subsurface biofilms) can generate colloids via detachment.
- More fundamentally, any proposed manipulation of the subsurface environment must be viewed as complex set of coupled interactions. The final outcome of any manipulation will reflex the integrated result of geochemical, hydrological, and microbiological processes.

The basic processes that fall under this topic were identified along with recommended research. A compilation of the discussion in outline form is given below:

APPROACH: DEVELOPMENT OF BIOTRANSFORMATION PROCESSES IN THE SUBSURFACE

Description

Biotransformation processes may be used to destroy contaminants. This could involve the introduction of microorganisms (involving processes controlling colloidal transport) or chemical modification of the groundwater environment to stimulate indigenous microorganisms. In addition to the assumed goal of stimulating contaminant degradation, these processes are also relevant to understanding biofilm development and the effects of biofouling and bioplugging on the permeability of a formation (see Session II). Growth of microorganisms may also release biocolloids to the bulk liquid flow, which could contribute to "log-jamming," as well as to "seeding" of downgradient regions of the formation. More fundamentally, biodegradation is but one aspect of the broader question of processes controlling the structure and function of microbial communities in subsurface environments.

Processes

Several processes were identified which need to be considered in conceptualization, and possible manipulation, of subsurface microbial function:

- *Bulk Liquid Transport.* Direct contact between contaminants and microorganisms is essential for biodegradation. Chemicals (growth substrates, nutrients, and electron acceptors) and microorganisms must be distributed via the bulk liquid ground water flow.
- *Attachment.* To be effective, microorganisms must accumulate on surfaces in the zone of contamination. The cell approaches a surface by transport. Attachment is the process of a cell associating with a surface and sticking to it.
- *Detachment.* This process is the loss of cells from the surface due to grazing, erosion, abrasion, and sloughing. The rate of detachment influences the time to accumulate biomass and to achieve a given biotransformation.
- *Growth Kinetics.* Growth kinetics describes the rate of metabolic activity of the accumulated microorganisms. This information describes how fast the microorganisms are utilizing contaminants and how fast the biomass is growing. The rate of decay of the microbial populations is also included.

Advantages

There is an urgent need to clean-up contaminated groundwater and associated soils and aquifer sediments. Clean-up methods often employ flushing the subsurface with water so that contaminants can dissolve and be pumped to the surface for above-ground treatment. Because organic contaminants generally sorb to soils, they are not readily leached from soils, and such above-ground treatment systems are generally inefficient and slow. Furthermore, most above-ground technologies involve physical/chemical processes that simply sequester the contaminants or transfer them to another environmental medium.

Many of the most serious contamination problems, particularly at DOE sites, involve contamination of deeper aquifers that are inaccessible to some of the most common above-ground treatment technologies. Development of in situ biotransformation processes in the subsurface offers a potentially more effective and economical cleanup technique through partial or complete destruction of the contaminants. The general approach in this process (termed bioremediation) is to stimulate the growth of indigenous or introduced microorganisms in regions of subsurface contamination, and thus, provides direct contact between microorganisms and the dissolved and sorbed contaminants for biotransformation. Effective manipulation of microbial transport and attachment are critical to the success of bioremediation.

Limitations

Use of bioremediation for treatment of subsurface regions is challenging because these regions are difficult to characterize and the introduction of chemicals and microorganisms is not easy. At present, there appear to be few options for modifying the transport and attachment characteristics of microorganisms; however, it was noted new information about the factors controlling biocolloid mobility may suggest ways of selecting or genetically engineering microorganisms optimized with respect to both metabolic and transport properties. There is, however, a concern over introducing new microbial species in the subsurface, particularly in the case of genetically engineered microorganisms. There is also concern over possible harmful residuals that remain in the subsurface following bioremediation. Coupling our poor knowledge of microbial transport and attachment with our limited understanding of factors controlling biotransformation pathways and reaction rates of many organic contaminants of concern makes establishing the utility of *in-situ* bioremediation an important scientific and engineering problem.

Gaps in Knowledge

Bioremediation provided a theme that integrated many of the knowledge gaps identified in other discussion sessions. Key areas in which more research is needed include:

- *Bulk Liquid Transport*
  - Research should be conducted to determine the relative contribution of advection/dispersion versus motility to microbial transport.
  - Better information is needed on the buoyant density of microorganisms. A broad survey of microorganisms should be conducted to compile buoyant densities that can be used to model colloidal transport. Can the buoyant density be controlled?
  - One or more studies should evaluate the relative dispersion of microorganisms in different geologic media at a variety of scales. The possible role of motility on dispersive characteristics is poorly understood.
- *Attachment*
  - Research should be conducted to determine the effect of cell surface properties on attachment.
  - Studies to evaluate the effect of environmental conditions on cell surface properties need to be carried out.
  - Better information is needed on the influence of deposited cells on subsequent deposition of microorganisms.
- *Detachment*
  - The factors affecting reversibility under steady-state conditions and in the presence of chemical and physical perturbations need to be identified.
  - Better information is needed on the rheology of biofilms.
  - Research should be conducted to determine the relationship between cell surface interactions (e.g. growth state and direction of growth) and detachment.
- *Growth Kinetics*
  - Procedures should be developed to determine the minimum inoculation size required to achieve a given bioremediation goal, such as establishment of a desired microbial population and contaminant clean-up.
  - Information is needed on how to activate cells and stimulate growth.
  - The impact of microbial manipulation on the microbial ecology (e.g., residuals and species displacement) needs to be determined.
  - Methods need to be developed to determine in-situ kinetics for microbial activity.
  - Methods need to be developed to monitor the performance of manipulated biotransformations in the subsurface.
  - Information is needed on the significance of grazing on the ability to accumulate subsurface biomass for remediation.
  - The survival of introduced species and the maintenance of viability needs to be determined.

DISCUSSIONS AND CONCLUSIONS

Our knowledge of colloidal behavior in groundwater systems is largely derived from studies of non-biological particles. Much of the group discussion was centered on the extent to which these concepts apply to biological particles. It was not difficult to generate a long list of research needs because the interest in microbial transport and attachment is recent and much remains to be learned. Microorganisms have diverse morphological features in comparison to non-biological particles. The metabolic activity of microorganisms also can alter the chemical character of the cell surface as well as the surrounding medium. These physical and chemical features greatly complicate the modelling of microbial transport and attachment. Most of the group members agreed that models for bulk liquid transport of non-biological particles are likely to be applicable for bulk liquid microbial transport.

There was strong consensus among the group that factors influencing detachment are largely unknown and better information on detachment is urgently needed. A few of the group members felt that solution chemistry is very important in most cases of microbial attachment. The group strongly supported the approach of using microorganisms for control of organic contamination in the subsurface.

Processes of bulk liquid transport, attachment, detachment, and microbial kinetics need to be coupled to describe the movement of biocolloids. Well integrated laboratory and field investigations are needed to advance our knowledge and address the knowledge gaps identified.

TABLE 15.1 Evaluation of strategies considered in group discussions.

PROPOSED MANIPULATION APPROACH	KEY HYDRO-GEO-BIO-CHEMICAL PROCESSES IN APPROACH	ADVANTAGES/REASONABLENESS OF APPROACH	LIMITATIONS/IMPRACTICALITIES OF APPROACH	MAJOR GAPS IN KNOWLEDGE	OVERALL FEASIBILITY
Maximize mobility of microorganisms	Transport Adhesion/attachment Detachment	New metabolic capabilities Utilization of sorbed compounds	Release of undesirable microorganisms	Growth state vs attachment "Alpha" values for bacteria Competition and interaction with other colloids	
Facilitate accumulation of biofilm	Biofilm kinetics Substrate availability Transport	Increased rates Clogging of large pores	Permeability reduction Nutrient limitation Small scale length	Intrinsic reaction rates Biodegradation of mixtures Detachment	
Remove accumulated biofilm	Biofilm kinetics/decay Shear/transport Surface chemistry of aquifer solids	Remove biofouling	Chemical controls (secondary effects of added chemicals)	Factors influencing detachment Rheology of biofilms	
Minimize transport of pathogens	Pathogen source Inactivation Attachment/deposition Control using phages	Prevent contamination Reduce buffer zone	Secondary effects with chemical manipulation	Pathogen survival Competitive displacement How low is "alpha"? Modeling	

SECTION III

MANIPULATING COLLOIDS TO CHANGE AQUIFER PERMEABILITY

CHAPTER 16

Overview

James R. Hunt

The transport of particles and colloids through clean porous media has received considerable attention, but such results are not predictive of conditions encountered in soils or aquifers. In natural systems the solid surfaces are not clean but are coated with previously deposited particles or precipitates formed in situ. These attached particles project into the pore space and can substantially alter medium permeability and thus fluid flow through and around such regions. The amount of particulate matter that can be accumulated with a porous region is unknown due to a complex interaction between hydrodynamics and the aggregate structure controlled by physical and chemical factors. Tien and Payatakes (1979) have reviewed deep bed filtration theories and McDowell-Boyer et al. (1986) provided an overview of these issues and the status of predictive models with an emphasis on contaminant transport issues in soils and aquifers.

Filtration theory for the removal of suspended particles from air and water has undergone considerable development over the past thirty years. Particle removal by the completely physical process of size exclusion is referred to as straining and usually results in a surface cake of retained particles. A deep bed network model for straining was presented by Rege and Fogler (1987, 1988) and showed good agreement with experimental data collected on clay accumulation in sandstone cores. Complete blocking of the sandstone occurred within the first few centimeters of the entry face of the core. If the suspended particle size is relatively small, the resulting deposit will have a low permeability. Straining or cake filtration is used industrially to concentrate solids from dilute suspensions, but frequent cake removal is required to achieve fluid flow. Particles much smaller than the medium size can be removed only if the particle collides and then sticks with a fixed solid surface. There are three dominant collision mechanisms for particles suspended in water that are highly dependent on particle size. For particles less than about 1 μm the dominant transport mechanism to fixed surfaces is Brownian motion with smaller particles having greater diffusivities than larger particles. For particles larger than 10 µm the dominant transport mechanisms are gravitational settling and size-based interception. Both settling and interception are dependent on particle size squared and thus these removal mechanisms rapidly increase with particle size. The size dependencies of these three mechanisms result in particles over the size range of 1 to 10 µm having the least number of collisions and the greater potential for migration in water-saturated porous media. These predictions were confirmed by the experimental work of Yao et al. (1970).

Particle collisions with fixed surfaces result in attachment only if physical and chemical interactions are favorable. The dominant force hindering attachment is the electrostatic repulsion between particles and surfaces since both are usually negatively charged in natural waters. If the solution electrolyte is sufficiently concentrated to mask electrostatic repulsion, then the attractive, but short-range, van der Waals force can cause particle attachment. High fluid shear can hinder particle attachment and steric forces associated with polymeric material at particle surfaces can help or hinder attachment. Predictions of particle attachment are very sensitive to surface potentials and solution composition. Experimental data reveal that attachment efficiency is far less sensitive to parameters than theory predicts, and far greater attachment rates are observed than are predicted when electrostatic forces are present. Tobiason and O'Melia (1988) and Tobiason (1989) have shown experimentally that particle removal by interception and settling had a minimum collision efficiency of 0.03 under clean bed conditions. Brownian removal has been studied by a number of researchers and Elimelech and O'Melia (1990) provide an updated review and new experimental results demonstrating the importance of physical-chemical interactions in controlling particle attachment even though available theories are not successful in predicting such results. Measured attachment efficiencies were always greater than 4×10^{-3}. These studies have not considered in detail the influence of retained particles on subsequent particle capture and permeability reduction. Darby and Lawler (1990) have measured particle capture and the increase in head loss for conditions of favorable attachment. They observed that particle accumulation also resulted in erosion of particle aggregates that were previously retained on fixed surfaces.

The studies on clean filters indicate that particle deposition is expected to occur over distances of meters under groundwater flow conditions, and the particles will accumulate as a deposit on the fixed surfaces. The particle deposit alters permeability, head loss, and particle dynamics. Such processes should be expected under conditions encountered in aquifers, but aquifers differ from water filters because aquifers are not homogenous or isotropic, and flow is driven by regional and local hydraulic gradients that change with time. Further consideration of particle deposition and permeability alteration in porous media requires consideration of fluid flow. Fluid flow is modeled by Darcy's law

$$v = Ki \tag{1}$$

where v is the approach velocity or flow rate divided by the total cross sectional area, i is the hydraulic gradient, -dh/dx, and K is the hydraulic conductivity that is a function of the fluid and medium properties

$$K = \frac{gk}{v} \tag{2}$$

where g is the gravitational acceleration, v is the kinematic fluid viscosity, and k is the medium permeability modeled by Carman and Kozeny as

$$k = \frac{d_m^2}{216} \frac{n^3}{(1-n)^2} \tag{3}$$

with d_m being the medium diameter and n as the porosity. Darcy's law is used in filtration and groundwater flow modeling, but the approach and units are different. For example, a porous media composed of 0.5 mm diameter sand with a porosity of 0.4 has a permeability $k = 2.1 \times 10^{-6}$ cm^2 or 200 darcy (1 darcy is approximately equal to 10^{-8} cm^2), and this permeability corresponds to a hydraulic conductivity of 0.2 cm s^{-1} or 170 m d^{-1}. Alternatively, for a pore size of 1 μm corresponding to a medium size of 7 μm, the permeability is 4.0×10^{-10} cm^2, or 0.04 darcy with a hydraulic conductivity of 4×10^{-5} cm s^{-1} or 0.04 m d^{-1}. At permeabilities less than 1 darcy fluid flow and hydraulic gradients less than 0.01, then fluid flow velocities are less than 0.01 m d^{-1}, and the pore size is small enough to strain out particles, colloids, and bacteria. Little regional contaminant migration is expected and any suspended particles should have limited migration.

As particles accumulate within the pore spaces, the pore sizes decrease along with the permeability. A direct application of equation (3) during particle retention is not successful, instead, various empirical expressions have been proposed as reviewed in McDowell-Boyer et al. (1986). The expressions that appear in the literature relating particle deposit with permeability reduction and particle capture are able to fit experimental data but are not predictive under new situations.

The critical issue in clogging is the interaction of fluid flow with the retained particles. The condition that will control deposition and erosion of particles is the fluid shear stress imposed at the interface between the fluid and the retained particles. Shear stress, τ, is given by

$$\tau = \rho v G \qquad (4)$$

where ρ is the fluid density and G is the velocity gradient, dU_x/dy, at the interface. In studies of cohesive sediment erosion by water currents in rivers, lakes and estuaries, a shear stress above some critical value is observed to cause erosion (Parchure and Mehta, 1985). For kaolinite clay particles the critical shear stress was in the range of 1 to 5 dyne cm^{-2}. While cohesive sediments in surface waters are not identical to particle deposits in porous media, the deposit structure is still controlled by physical-chemical interactions between micrometer-sized particles in very open and fragile structures. This concept of a critical shear stress or aggregate strength suggests conditions for maintenance of flow through porous media containing deposits.

For a deposit having a critical shear stress, τ_c, over the range of 1 to 10 dyne cm^{-2}, it is possible to estimate fluid flow conditions required to maintain flow in porous media containing such a deposit. The critical velocity gradient or fluid shear rate at the fluid-deposit interface from equation (4) is given by

$$G_c = \frac{\tau_c}{\rho v} \qquad (5)$$

which takes on values of 100 to 1000 s^{-1}. This velocity gradient can be related to the hydraulic gradient, i, by recognizing that fluid shear causes energy dissipation that is reflected in head loss. The relationship was presented by Ives (1970) for filters as

$$G_c \approx \left(\frac{g v i}{n v} \right)^{1/2} \qquad (6)$$

The expression was derived for clean porous media and is only approximate because the porosity is reduced by retained particles in an unknown way. Using Darcy's law for v and expressing the hydraulic conductivity in terms of permeability using equation (2) results in

$$\frac{\tau_c}{\rho \, v} \approx \frac{g}{v} \left(\frac{k}{n} \right)^{1/2} i_c \qquad (7)$$

or

$$i_c \approx \frac{\tau_c}{\rho \, g} \left(\frac{n}{k} \right)^{1/2} \qquad (8)$$

A similar criterion for erodability was presented by Khilar et al. (1985).

Equation (8) establishes a relationship between permeability and the necessary hydraulic gradient to maintain flow. If a clogging formation has a reduced permeability of 100 darcy, or 10^{-6} cm^2, a porosity of 0.4, and a critical shear stress of 1.0 dyne cm^{-2}, then a critical hydraulic gradient of 0.65 is necessary to achieve a balance between deposition and erosion. Higher hydraulic gradients will erode deposited material causing an increase in permeability. For reduced hydraulic gradients, further deposits will accumulate until the formation clogs completely. Thus if there exists a critical shear stress for the deposited particles, then equation (8) defines the resulting equilibrium permeability, k, for a fixed hydraulic gradient. If the equilibrium permeability is greater than the clean media permeability, then no deposition will occur.

Experience with porous media clogging is available in the water filtration field and in petroleum reservoir engineering. In both cases hydraulic gradients exceeding 1.0 are typically observed under constant flow and constant pressure conditions. Permeability reductions to 0.1 to 1% of the original permeability are observed under constant flow conditions and complete clogging is expected under constant pressure conditions if columns are sufficiently long.

For fluid flow in soils and aquifers, the available hydraulic gradients are insufficient to maintain flow if particle suspensions are present that attach to fixed surfaces. The accumulated experience on clogging in filters and sandstones report hydraulic gradients typically exceeding 1.0. Regional hydraulic gradients in aquifers rarely exceed 0.01 although gradients near one are possible near well screens and during vertical infiltration in the unsaturated zone. Given a sufficient supply of suspended particles entering an aquifer then complete clogging should be expected. The retained particles can be remobilized by increasing the hydrodynamic force on the deposit, or by chemically altering the physical-chemical forces holding the deposit together.

This overview has considered only inorganic mineral deposits and not microbially active biofilms. Biofilms will tend to grow and block pores as long as substrates are supplied. Biofilms also have strength characterized by a critical shear stress and erosion or sloughing will occur if hydrodynamic forcing exceeds the biofilm strength. If flow is reduced by decreasing permeability, then further growth is limited and components of the biofilm would decay and maintain a steady state permeability as observed by Cunningham in this section.

ACKNOWLEDGMENT

Participation at the meeting was supported by Contract W-7405-ENG-48 from the Department of Energy to the Lawrence Livermore National Laboratory.

REFERENCES

Darby, J.L. and Lawler, D.F., Ripening in depth filters: effect of particle size on removal and head loss, *Environ. Sci. Technol.*, 24(7):1069-79, 1990.

Elimelech, E. and O'Melia, C.R., Kinetics of deposition of colloidal particles in porous media, *Environ. Sci. Technol.*, 24(10):1528-36, 1990.

Ives, K.J., Rapid filtration, *Water Res.*, 4:201-23, 1970.

Khilar, K.C., Fogler, H.S., and Gray, D.H., Model for piping-plugging in earthen structures, *J. Geotechn. Eng.*, 111(7):833-47, 1985.

McDowell-Boyer, L.M., Hunt, J.R., and Sitar, N., Particle transport through porous media, *Water Resour. Res.*, 27(4):665, 1991.

Parchure, T.M. and Mehta, A.J., Erosion of soft cohesive sediment, *J. Hydraulic Eng.*, 111(10):1308-26, 1985.

Rege, S.D. and Fogler, H.S., Network model for straining dominated particle entrapment in porous media, *Chem. Eng. Sci.*, 42(7):1553-64, 1987.

Rege, S.D. and Fogler, H.S., A network model for deep bed filtration of solid particles and emulsion drops, *AIChE J.*, 34(11):1761-72, 1988.

Tien, C. and Payatakes, A.C., Advances in deep bed filtration, *AIChE J.*, 25(5):737-59, 1979.

Tobiason, J.E., Chemical effects on the deposition of non-Brownian particles, *Colloids and Surfaces*, 39:53-77, 1989.

Tobiason, J.E. and O'Melia, C.R., Physicochemical aspects of particle removal in depth filtration, *J. Amer. Water Works Assoc.*, 80(12):54-64, 1988.

CHAPTER 17

Influence of Biofilm Accumulation on Porous Media Hydrodynamic Properties

A.B. Cunningham

In porous media, as in other aqueous environments, microbial cells may exist in suspension or adsorb firmly to solid surfaces comprising the effective pore space. If favorable environmental conditions persist, adsorbed cells will grow and reproduce at the surface, increasing the amount of attached biomass. Escher (1986) observed that, under conditions of constant nutrient flux and laminar flow in capillary tubes, sorption-related processes were governed by suspended cell concentration, whereas the growth process at the surface was a function of microbial surface concentration.

If rates of cell adsorption and growth exceed the rate of desorption, a net accumulation of biomass will result on the surface. As the accumulation process continues, additional cells may attach (and detach) directly to (and from) the existing biomass surface. Attachment and detachment are probably the least understood processes affecting the accumulation of biofilm. Trulear and Characklis (1982) and Chang and Rittman (1988) used a first-order expression to model net detachment rate (i.e., detachment rate minus attachment rate) as a function biomass. Speiter and DiGiano (1987) proposed adding a growth-related detachment rate term to the first-order biomass model to account for high detachment rates observed during periods of high cell growth.

Particles of organic and inorganic material flowing in suspension may be removed by the attached biomass through filtration processes including diffusion, interception, and sedimentation (Bouwer 1987). The entire deposit of cells and polymers, together with captured organic and inorganic particles, is termed the "biofilm." The amount of biofilm accumulation occurring in a porous media flow system is therefore the net result of the

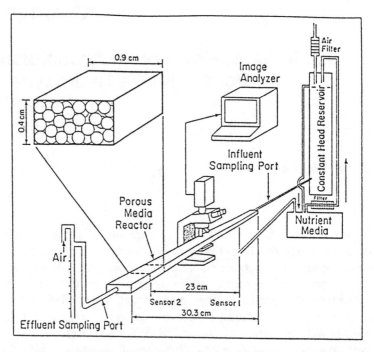

Figure 1. Experimental system for monitoring biofilm accumulation along a rectangular porous media reactor.

biomass added through adsorption, growth, attachment, and filtration, less the amount removed by desorption and detachment.

Individual biofilm processes are considerably more difficult to examine in porous media than in other common reactor geometries such as flasks, tanks, reservoirs and pipelines. Biofilm growth, for example, is complicated by the nature of fluid and nutrient transport which, in porous media, occurs along tortuous flow paths of variable geometry. Similarly, the wide distribution of pore velocities introduces considerable variation in the processes of adsorption, desorption, attachment, and detachment.

Laboratory-scale porous media biofilm reactors (Figure 1) were used to evaluate the effect of biofilm accumulation, measured as the average thickness along a 50 mm flow path, on media porosity, permeability, and friction factor. Media tested consisted of 1-mm glass spheres, 0.70-mm sand, 0.54-mm sand, and 0.12-mm glass and sand. *Pseudomonas aeruginosa* was used as inoculum, and 25 mg L^{-1} glucose substrate was continuously supplied to the reactor. Reactors were operated under constant piezometric head conditions, resulting in a flow rate decrease as biofilm developed.

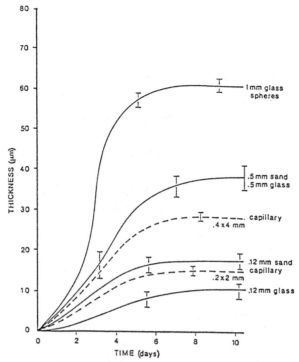

Figure 2. Progression of biofilm thickness (*Ps. aeruginosa*) for media of different diameter and composition. All reactors were run in parallel under a piezometric gradient of 0.5.

The progression of biofilm thickness for porous media of various sizes and composition were generated by simultaneously running five parallel reactors under a constant piezometric head gradient of 0.5 cm/cm (Figure 2). Comparison of these accumulation curves indicates that the ultimate biofilm thickness varied directly with media pore space size and was not influenced significantly by media composition. A similar observation was made using "empty" rectangular reactors (i.e., containing no porous media) of dimension 0.2 × 4 mm on 0.4 × 4 mm; the reactor with the larger cross section (0.4 × 4 mm) and, hence, the larger velocity exhibited a greater maximum biofilm thickness. In every case the progression of biofilm thickness followed a sigmoidal shaped curve reaching a maximum thickness after about 5 days.

Media porosity decreased between 50 and 96% (Figure 2) with increased biofilm accumulation, while permeability decreased between 92 and 98% (Figure 4). Porous media friction factor increased substantially for all media tested. Observations of permeability in the biofilm-media matrix indicate that a minimum permeability (3 to 7 × 10^{-8}mm^2) persisted after

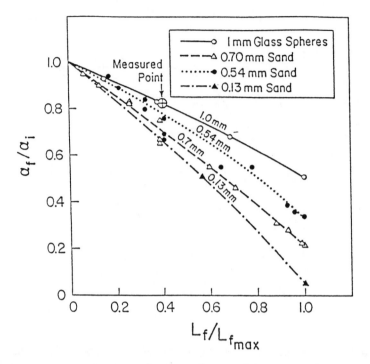

Figure 3. Variation in estimated media porosity, α_f with biofilm thickness. For 1 mm glass spheres $\alpha_{max} = 0.48$, $L_{f\,max} = 65$ μm; for 0.7 mm sand $\alpha_{max} = 0.35$, $L_{f\,max} = 15.4$ μm; for 0.54 mm sand $\alpha_{max} = 0.35$, $L_{f\,max} = 46$ μm; for 0.12 mm sand $\alpha_{max} = 0.47$, $L_{f\,max} = 15.3$ μm. The $\otimes$ symbol represents measured $\alpha_f(.34)$ and L_f (25 μm) for 1 mm glass spheres.

biofilm thickness has reached a maximum value. Such results indicate substantial interaction between mass transport, hydrodynamics, and biofilm accumulation at the fluid-biofilm interface in porous media.

As biofilm thickness increases, the diffusional path length within the biofilm increases, thereby decreasing nutrient concentrations in the base-film. Providing the piezometric head gradient remains constant, increased thickness will also result in decreased pore velocity. Decreased pore velocities will reduce both advective and dispersive transport, thereby lowering nutrient concentrations at the film-water interface, subsequently reducing growth rate. Decreased pore velocities will also reduce shear stress, thereby reducing the rate of detachment. Accumulation of biofilm will continue until specific growth rate is balanced by detachment rate. These

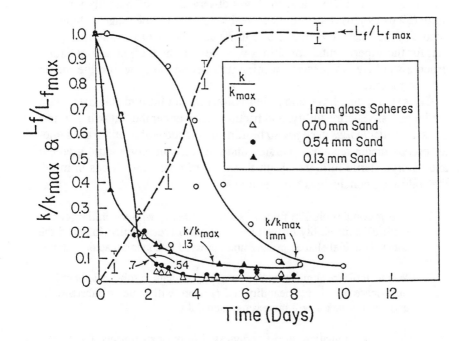

Figure 4. Porous media permeability decrease corresponding to increased biofilm thickness. Lf/Lf$_{max}$ is a single composite dimensionless curve representing all thickness curves from Figure 5. K$_{max}$ values are 2.1 x 10^{-5} cm^2 (1 mm glass spheres), 3.19 x 10^{-6} cm^2 (0.70 mm sand), 2.17 x 10^{-6} cm^2 (0.54 mm sand) and 9.7 x 10^{-7} cm^2 (0.12 mm sand).

interactions give rise to the sigmoidal shape exhibited by the accumulation progressions (Figure 2), which indicate a quasi-steady-state thickness is reached after about 5 days of reactor operation. Maximum biofilm thickness values of 63 µm, 40 µm, and 9 to 14 µm, were observed (Figure 2) for media particle diameters of 1 mm, 0.54 mm, and 0.12 mm. Clean-surface permeability values of 2.1 × 10^{-5}, 2.17 × 10^{-6}, and 9.7 × 10^{-7} cm^2 correspond to these particle diameters, suggesting a direct relationship between media permeability and maximum biofilm thickness.

After biofilm thickness reaches quasi-steady state, the volume of effective pore space (permeability) appears to stabilize and remain constant until operating conditions are disturbed. The media/biofilm permeability (for media of various size and composition) stabilized and remained essentially constant after about 5 days of reactor operation (Figure 4). Torbati et al.

(1986) has reported a similar observation wherein in situ accumulation of microorganisms in Brea sandstone was observed to preferentially plug the large (59 to 69 μm) pore entrances, resulting in a narrow range of biofilm-encased pore sizes in the range of 5 to 30 μm. These observations suggest that for the experimental conditions used herein, the biofilm accumulation process stabilizes to preserve a minimum permeability within the media biofilm matrix.

Our experiments have correlated porous media biofilm thickness with media porosity, permeability, and friction factor under the following conditions: a *Pseudomonas aeruginosa* biofilm developed within uniform-diameter porous media reactors, operated under a constant piezometric gradient and using a sterile influent. With these experimental conditions in mind, the following conclusions can be drawn:

1. The maximum biofilm thickness varied directly with the initial (clean-surface) permeability of the media. Biofilm accumulation followed the same sigmoidal-shaped progression observed in conduit reactor systems.

2. When biofilm accumulation becomes large enough to substantially reduce pore space, media permeability and porosity will decrease substantially and friction factor will increase substantially.

3. As the accumulation process progressed in our experiments, the permeability of the biofilm-media matrix stabilized at a minimum value (3 to $7 \times 10^{-8} cm^2$), regardless of media particle diameter.

4. Predictions of formation plugging and biofilm accumulation in porous media can be accomplished based on biofilm kinetics and the concept of theoretical porosity.

ACKNOWLEDGMENTS

Permission to reprint Figures 1-4 has been granted by the American Chemical Society, September 23, 1991.

The author acknowledges support from the Center for Interfacial Microbial Process Engineering at Montana State University, a National Science Foundation Engineering Research Center and the Center's Industrial Associates. Although the research described in this article has been funded in part by the U.S. Environmental Protection Agency under Assistance Agreement R-815709 to Montana State University through the Hazardous Substance Research Center for U.S. EPA regions 7 and 8 headquartered at Kansas State University, it has not been subjected to the Agency's peer and administrative review and, therefore, may not necessarily reflect the views of the Agency, and no official endorsement should be inferred. This research was also supported by the U.S. Geological Survey (Project 14-08-0001-G1284).

REFERENCES

Bouwer, E.J., Theoretical investigation of particle deposition in biofilm systems, *Water Res.*, 21(12):1489-98, 1987.

Chang, H.T. and Rittman, B.E., Comparative study of biofilm shear loss on different adsorptive media, *J. Water Pollut. Control Fed.*, 60(3):362-68, 1988.

Escher, A.R., Bacterial colonization of a smooth surface: An analysis with image analyzer, Ph.D. dissertation, Montana State University, 1986.

Speitel, G.E., Jr. and DiGiano, F.A., Biofilm shearing under dynamic conditions, *J. Environ. Eng.*, 113(3):464-75, 1987.

Torbati, H.M., Raiders, R.A., Donaldson, E.C., McInerney, M.J., Jenneman, G.E., and Knapp, R.M., Effect of microbial growth on pore entrance size distribution in sandstone cores, *J. Ind. Microbiol.*, December: 227-34, 1986.

Trulear, M.G. and Characklis, W.G., Dynamics of biofilm processes, *J. Water Pollut. Control Fed.*, 54(9):1288-1301, 1982.

CHAPTER 18

Biomass Manipulation to Control Pore Clogging in Aquifers

Peter R. Jaffe and Stewart W. Taylor

Changes in growth substrate and pressure head were monitored over time and space for a one year period in two 50-cm-sand-(ASTM C-190) filled laboratory columns to which a low-concentration growth substrate was fed continuously at constant flow rates (Jaffe et al. 1988; Taylor and Jaffe 1990a). The concentration of growth substrate and flow applied to each column differed so that the total mass of growth substrate per unit time applied to column 1 was ~3.9 times that applied to column 2. The substrate concentration was sufficiently low to maintain an aerobic effluent in both columns. Changes in permeability were calculated from the changes in the head and are shown in Figure 1 as a function of time and space. Tracer studies were conducted toward the end of the experiment to determine spatial changes in dispersivity (Jaffe et al. 1988; Taylor and Jaffe 1990b). At the end of the experiment, the columns were dismantled and the spatial distribution of biomass quantified (Figure 2). Changes in porosity, permeability, and dispersivity in the porous media, from the formation of a biofilm, were described successfully with a cut-and-random-rejoin-type model (Mualem 1976; Taylor et al. 1990). For this purpose it was assumed that a uniform biofilm thickness develops in the individual pores. Based on the initial pore size distribution and how it is modified by a uniform biofilm thickness, a new pore size distribution can be computed and, hence, a new porosity, permeability, and dispersivity. Mass balance equations for the attached and suspended biomass and for growth substrate were written and coupled with the pore-structure model (Taylor and Jaffe 1990c). This model was verified against the column data, and used to examine the relation between soil properties (grain-size distribution, porosity, and permeability) and clogging propensity of the soil resulting from biomass growth. Results from a sensitivity analysis showed that the largest decrease in relative permeability can be expected for soils with a wide

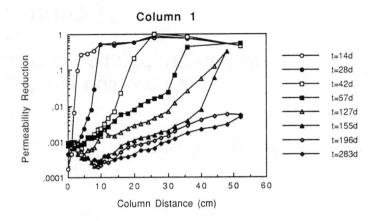

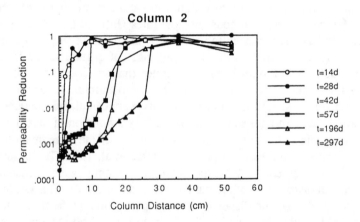

Figure 1. Permeability reduction (K/K_o) as a function of time and space.

pore-size distribution. In general, uniform soils with a pore size distribution index larger than two are not very likely to be significantly affected by the development of a biofilm. The model was also used to investigate biomanipulation strategies that minimize soil clogging (Taylor and Jaffe unpublished data). The most severe soil clogging resulting from biomass growth is observed when oxygen and a growth substrate are injected simultaneously through the same well. For this case, given the magnitude of biokinetic rates and flow velocities in the vicinity of injection wells, sufficient biomass to affect the soil's permeability only develops in the direct vicinity (a few feet) of the well head. Mitigation of well plugging by injecting the oxygen and growth substrate out of phase was investigated

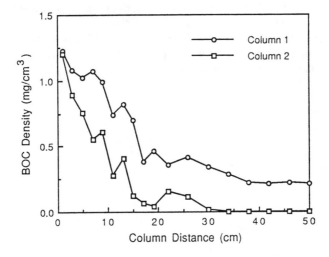

Figure 2. Biomass buildup in the columns as organic carbon: after 284 days of operation for column 1 and 356 days of operations for column 2.

(Taylor and Jaffe unpublished data). Results from mathematical simulations showed that if the same mass of oxygen and growth substrate are injected, in one case simultaneously and in another case out of phase, the buildup of head was reduced by an order of magnitude. No significant reduction in the soil's permeability was predicted for the case where only oxygen is injected into a formation containing a growth substrate, even if the soil had a wide pore-size distribution.

REFERENCES

Jaffe, P.R., Taylor, S.W., Baek, N.H., Milly, P.C.D., and Marinucci, A.C., Biodegradation of trichloroethylene and biomanipulation of aquifers, DIOWRT, NTIS No. PB89 178594/AS, 1988.

Mualem, Y., A new model for predicting the hydraulic conductivity of unsaturated porous media, *Water Resour. Res.*, 12(3):513-22, 1976.

Taylor, S.W. and Jaffe, P.R., Biofilm growth and the related changes in physical properties of a porous media, 1. Experimental investigation, *Water Resour. Res.*, 26(9):2153-59, 1990a.

Taylor, S.W. and Jaffe, P.R., Biofilm growth and the related changes in physical properties of a porous media. 3.- Dispersivity, *Water Resour. Res.*, 26(9):2171-80, 1990b.

Taylor, S.W. and Jaffe, P.R., Substrate and biomass transport in a porous medium, *Water Resour. Res.*, 26(9):2181-94, 1990c.

Taylor, S.W., Milly, P.C.D., and Jaffe, P.R., Biofilm growth and the related changes in physical properties of a porous media. 2. Permeability, *Water Resour. Res.*, 26(9):2161-69, 1990.

CHAPTER 19

Colloid Entrainment, Transport, and Entrapment in Two Phase Flow in Porous Media

M.M. Sharma

The effect of two flowing fluid phases on colloid release, entrainment and trapping was studied using flat plate experiments, bead packed flow visualization micromodels and core flooding experiments. The experiments clearly show how various factors such as particle wettability, size and location on the collector surface affect their release. Interfacial instabilities, and saturation history are also shown to control the release of individual particles as well as the breakup of aggregates. The flow visualization observations can partly be explained by a simple analysis of the surface forces acting between the particle-collector-fluid interface system.

Several investigators (Mungan 1968; Mungan 1965; and Khilar and Fogler 1984) have shown the existence of a critical salt concentration (CSC) beyond which the permeability of a rock varies with the salinity. Performing numerous experiments with monovalent and divalent ions, Sharma et al. observed that cations that are specifically adsorbed on mineral surfaces prevent colloid migration. Adopting the DLVO theory, the onset of deposition and release of clays was predicted and found to adequately model previously reported experimental behavior.

Another important reported observation is the existence of a critical flow velocity above which fines are entrained in the flowing fluid (Gruesbeck and Collins 1982; Gabriel and Iamdar 1983; Chamoun et al. 1989). Critical velocities have been observed by a number of investigators (Sarkar et al. 1990; Sharma et al. 1987; Chamoun et al. 1989). The entrained particles are subsequently captured at pore throats, resulting in significant reductions in permeability (Sarkar et al. 1990; Sharma et al. 1987).

In the presence of two fluid phases, the entrainment and subsequent capture of colloids is influenced by additional factors such as wettability, fluid saturations, and mobilities (Muecke et al. 1979; Sarkar et al. 1990).

The flow visualization studies were conducted using a microscope attached to a video recorder, which is used to observe flow through a two-dimensional microcell. Two kinds of cells were used in this study, (1) a flat plate flow channel cell, and (2) a packed glass bead cell. The flow channel cell is constructed by placing together two microscope slides (1.5 × 3.0 in.) separated by a 0.5-mm-thick gasket. For porous media studies, glass beads, 100 µm in diameter are packed inside the gap in the cell. The bead pack so formed is only 3 to 4 bead diameters thick.

A Zeiss transmitting light microscope with a bright-field mode condenser was used to observe the flow. Polystyrene particles (sizes ranging from 0.9 to 5 µm), glass (sizes ranging from 2 to 5 µm) and kaolinite particles (< 0.5 µm) were used as model "fines."

In the flat-plate flow channel experiments, the particles were deposited in the aqueous phase and the oleic phase was allowed to displace it. A torque balance (Chamoun et al. 1989) was used to predict whether the particle would be displaced forward with the fluid-fluid interface or be engulfed by the nonwetting phase. These calculations depend on the surface charge properties of the fluid-fluid interface as well as those for the substrate and the deposited particle. A summary of the experimental observations made and the model predictions is provided in Table 1. The theory provides a general agreement with observations.

Table 2 represents a summary of the observations made in the two-phase flow experiments conducted in the glass bead cell. As indicated in Table 2, when particles are subjected to drainage, the movement of particles is restricted to the breakage of weak particle bridges and movement toward pendular rings. In the case of kaolinite clay particles, destabilization of bridges as the oil phase enters the pore throat was more frequent than with other types of particles. Coagulation and accumulation of particles in the pendular rings was observed in all the experiments. In pores where the residual water filled the pore body and oil did not penetrate, particles remained in their original positions. In pores filled with the oil phase, the particles were located in the pendular rings, inside the residual water phase.

During imbibition, particles that are located in the pendular rings are released into the water phase as displacement of oil from the pores occurs. Some particles engulfed by the oil phase tend to flow with the oil through the porous medium. Successive movement of decane rheons in the porous medium also causes particle engulfment as the newly released particles from pendular rings are suspended in the water and physically impair flow ahead of the oil.

Table 1. Summary of flat plate experimental results.

Case			Experimental	Theoretical
Case I		W wets S2 O wets S1 Drainage	S1 is engulfed by oil. Particles are in the oil phase.	S1 is engulfed by oil. Particles attracted toward the interface.
Case II		W wets S2 O wets S1 Imbibition	Particles mobilized into oil phase and flow with oil.	Movement of S1 with the mobile O. S1 remains in O.
Case III		W wets S2 W wets S1 Drainage	Movement of particles along the pore surface	Movement along the surface is more favorable than release from it.
Case IV		W wets S2 W wets S1 Imbibition	S1 flows with W. No particles in oil phase.	S1 remains in O or near the interface.

Table 2. Summary of results from the flow visualization micromodels.

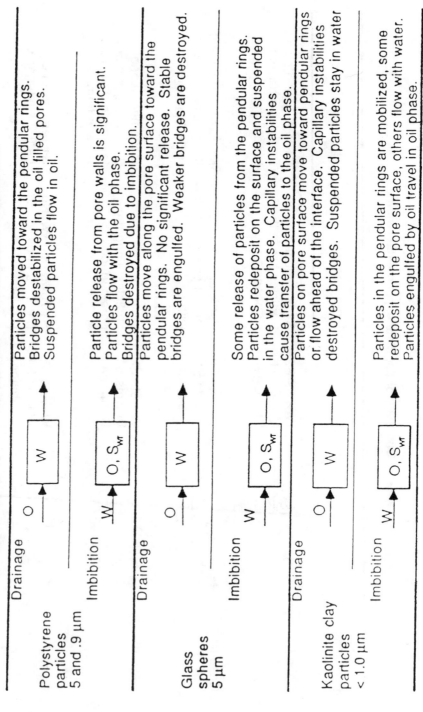

Polystyrene particles 5 and .9 μm	Drainage O → [W] →	Particles moved toward the pendular rings. Bridges destabilized in the oil filled pores. Suspended particles flow in oil.
	Imbibition W → [O, S_{WT}] →	Particle release from pore walls is significant. Particles flow with the oil phase. Bridges destroyed due to imbibition.
Glass spheres 5 μm	Drainage O → [W] →	Particles move along the pore surface toward the pendular rings. No significant release. Stable bridges are engulfed. Weaker bridges are destroyed.
	Imbibition W → [O, S_{WT}] →	Some release of particles from the pendular rings. Particles redeposit on the surface and suspended in the water phase. Capillary instabilities cause transfer of particles to the oil phase.
Kaolinite clay particles < 1.0 μm	Drainage O → [W] →	Particles on pore surface move toward pendular rings or flow ahead of the interface. Capillary instabilities destroyed bridges. Suspended particles stay in water
	Imbibition W → [O, S_{WT}] →	Particles in the pendular rings are mobilized, some redeposit on the pore surface, others flow with water. Particles engulfed by oil travel in oil phase.

Glass, kaolinite, and bentonite are generally considered to be water-wet particles. Their behavior in porous media experiments showed their tendency to remain in the water phase. On the other hand, oil-wet particles, as observed in the experiments, are strongly inclined to exist in the oleic-phase. Nowhere is this fact more obvious than in drainage experiments where oil-wet bentonite clay particles are engulfed and flow with the decane. Imbibition experiments also indicated no transfer of these particles from decane to water. Polystyrene particles could flow with either phase, depending on which phase contacted them first, indicating a neutral wettability. There is general agreement between model predictions and experimental results. Table 2 represents a summary of findings for both experimental and theoretical approaches.

Haines' jump phenomena (or rheons) can be described as any sporadic configurational instability that accompanies the displacement of a fluid interface from a pore. Successive occurrence of this phenomenon can affect the detachment and distribution of oil-wet and water-wet particles. High fluid velocities can be attained during the jump. Such a sporadic movement of the interface creates pressure waves that can destabilize particle bridges formed during deposition. Such behavior is consistently observed in experiments conducted with glass and kaolinite clay particles. In these experiments, particles and bridges 3 to 4 pores ahead of the fluid-fluid interface were detached as a result of local pressure fluctuations caused by the sudden movement of fluid interfaces. The released particles are less likely to be engulfed and transferred to the oil phase.

A detailed study was completed (Sarkar 1990) to investigate the permeability reduction in sandstone cores caused by colloid mobilization in the presence of two fluid phases. Large reductions in permeability were observed to occur in single-phase flow as the core was exposed to low salinity brines (Figure 1). Significantly smaller reductions in permeability occurred in the presence of a residual oil phase. Smaller permeability reductions were observed when the core was made oil-wet by the presence of the oil (Sarkar et al. 1990).

In single phase flow in porous media, low salt concentrations, high pH, and high flow rates cause the entrainment of colloidal particles. In the presence of two fluid phases, the following additional factors have been shown to strongly influence colloid entrainment, transport, and entrapment:

- particle and substrate wetting preferences,
- presence of residual phase saturations, and
- fractional flow of the two fluid phases.

The influence of each of these factors has been investigated through experiments in flat plate and glass bead micromodels and sandstone core studies.

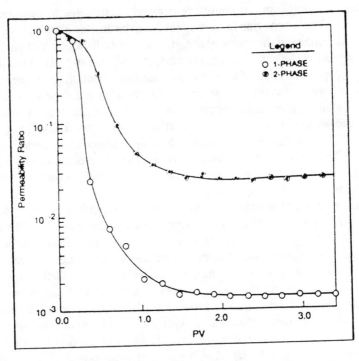

Figure 1. Comparison of single phase and two-phase permeability reductions in sandstone cores.

In soil-rock formations that contain substantial quantities of mobile fines, large reductions in permeability can be affected through the manipulation on in situ colloids. The extent of permeability reduction is,however, very sensitive to the type and quantity of fines and on the fluid phases in the pore space.

REFERENCES

Chamoun, H., Schechter, R.S., and Sharma, M.M., Hydrodynamic forces necessary to release non-Brownian particles attached to a surface, ACS Symposium Series 396, 1989, 548-59.

Gruesbeck, C. and Collins, R.E., Entrainment and deposition of fine particles in porous media, *Soc. Pet. Eng. J.,* 847-56 December 1982.

Gabriel, G.A. and Iamdar, G.R., An experimental investigation of fines migration in porous media, SPE Paper 12168, Presented at the 58th Annual Technical Conference, San Francisco, CA, 1983.

Khilar, K.C. and Fogler, J.S., The existence of a critical salt concentration for particle release, *J. Colloid Interface Sci.*, 101:214-24, 1984.

Muecke, T.W., Formation fines and factors controlling their movement in porous media, *J. Pet. Technol.*, 144-150 February 1979.

Mungan, N, Permeability reduction through changes in pH and salinity, *J. Pet Technol.*, 1449-53, December 1965.

Mungan, N., Permeability reduction due to salinity changes, *J. Canadian Pet. Technol.*, 20:113-17, 1986.

Sarkar, A.K. and Sharma, M.M., Fines migration in two-phase flow, *J. of Pet. Technol.*, 646-52, May 1990.

Sharma, M.M. and Yortsos, Y.C., Fines migration in porous media, *AIChE J.*, 33(10):1654-62, 1987.

Sharma, M.M., Yortsos, Y.C., and Handy, L.L., Release and deposition of clays in sandstones, SPE Paper 13562, Presented at the International Symposium on Oilfield and Geothermal Chemistry, 1985.

CHAPTER 20

Effects of pH on Fines Migration and Permeability Reduction

H. Scott Fogler and R.N. Vaidya

The release and migration of fine particles within a porous medium is a problem of great concern in the petroleum industry because it significantly influences the production of oil and gas from reservoirs. Our earlier studies in the area of fines migration have shown (1) the existence of a critical salt concentration above which the phenomenon of fines migration does not occur and (2) the effect of the rate of salinity change on the process of formation damage (i.e., reduction in permeability). A review on the subject of colloidally induced fines migration is published by Khilar and Fogler (1987). These earlier studies showed that the problem of fines migration is actually a problem in colloid chemistry and, therefore, needs to be addressed through the fundamental understanding of this science and it's application to the migration problem. Experimental observations also indicated that processes such as adsorption, ion-exchange, and dissolution can be important and must be considered in conjunction with the studies in colloidal chemistry. This phenomenon of a reduction in permeability, which is also known as sandstone water sensitivity, is observed to occur when fresh or low-salinity water replaces saltwater in sandstones. The result of a typical experiment to demonstrate this phenomenon is shown in Figure 1. In this experiment, a sandstone core saturated with 0.51 M NaCl solution is abruptly exposed to flow of freshwater (distilled water). The pressure drop across the core during brine and freshwater flow is monitored continuously as a function of time while maintaining the injection flow rate constant. The permeability is calculated using Darcy's law. We observe that the pressure drop increases drastically [or relative permeability (K/K_o) decreases drastically] after the injection of few pore volumes of freshwater.

These results have been explained on the basis of the DLVO theory. Initially, during the brine flow, the fine-clay particles are attached to the

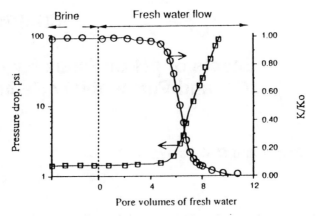

Figure 1. Permeability decline during a water shock (Vaidya and Fogler, 1990b).

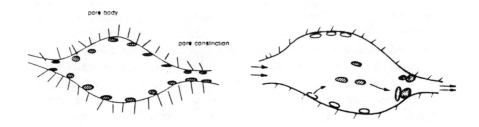

Figure 2. Pore throats before and after exposure to fresh water.

pore walls. However, when the flow is switched from saltwater to fresh-water, the electric double-layer expands, and the clay particles are released from the pore walls. These particles are carried by the permeating fluid until they are captured at pore throats, causing enormous reduction in per-meability. Figure 2 schematically delineates the mechanism for damage from fines migration. This mechanism also explains the recovery of per-meability during flow reversal.

The pH of distilled water used for these experiments was adjusted by using only HCl for pH values below 7 and only NaOH for pH above 7. When the brine-conditioned core is exposed to freshwater at pH 2.0 (zero NaCl concentration), no decline in permeability is observed, in agreement

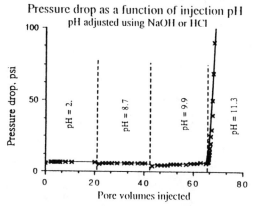

Figure 3. Pressure drop as a function of pH of the injection fluid (Vaidya and Fogler, 1990b).

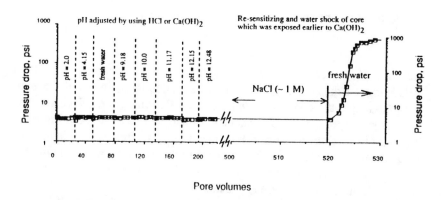

Figure 4. Pressure drop charges with change in pH using $Ca(OH)_2$.

with the results of Kia et al. (1987). However, when we continued to increase the pH of the injected fluid in step increments, a surprising result was observed as is shown in Figure 3: a rapid and drastic increase in the pressure drop, similar to that observed for a water shock, at a pH greater than 11.0. This implies that severe damage is caused on contact with the high-pH fluid when no salt is present. Data from similar experiments show comparable results.

To study the effect of other cations, other alkaline solutions such as KOH, NH_4OH, and $Ca(OH)_2$ were also used to change the pH of the injected fluid. Both K^+ and NH_4^+ showed results similar to those obtained

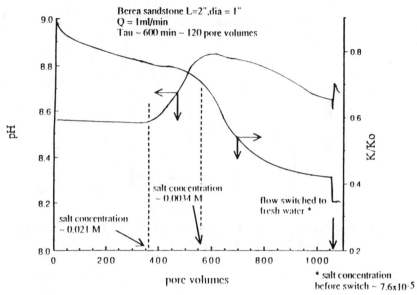

Figure 5. Permeability changes during gradual salinity decrease.

for Na⁺. However, when Ca^{+2} as used, no damage occurred even at pH values greater than 12 (Figure 4). To be sure that the core is still suscepti-ble to damage, the core was resensitized by flowing 1.0 M NaCl solution and subjected to a freshwater shock. As can be observed from Figure 4, a drastic reduction in permeability occurs during freshwater flow. The results were explained by using the DLVO theory in conjunction with the electrophoretic measurements (zeta potentials) of the involved surfaces (i.e., kaolinite clay and Berea sandstone).

Salinity-reduction experiments similar to those of Khilar (1981) were carried out. In these experiments, the salinity was reduced gradually by using a CSTR until the salt concentration had decreased *significantly below* the CSC (i.e., < 500 ppm). The result of a typical experiment is shown in Figure 5. Note that during the gradual decrease in salinity, the pH of the effluent did not show a significant increase (i.e., pH remains below 9.0). This result is in contrast to the results of a freshwater shock, where the effluent pH increased beyond 10.3 (Figure 6). Because there is no signifi-cant increase in pH during this time period, no drastic reduction in perme-ability occurs.

These results show that a combination of high pH and low salinity causes a release of fines in Berea sandstone, which leads to a drastic de-cline in the permeability of the medium. Measurement of the effluent pH

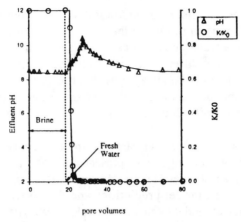

pore volumes

Figure 6. Permeability reduction and effluent pH transient during a water shock.

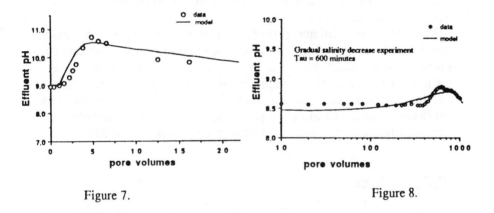

Figure 7.   Figure 8.

Comparison of model predictions for effluent pH transient with data.

during a water shock shows a transient increase, and results of a typical experiment are shown in Figure 6. These results and other key experiments established the interplay between salinity changes, cation exchange, and pH during a water shock and elucidated the vital role of the ion-exchange process in formation damage. An important conclusion of these findings was that the increase in pH accompanying the salinity change is the "triggering" mechanism initiating the release of fines. This was also verified by Vaidya and Fogler (1990b) by subjecting a brine saturated core to a water shock in the presence of a acid buffer.

The answer, therefore, to controlling damage from colloidally induced fines migration is to minimize the enhancement of the DLVO repulsive forces during a water shock. It can be achieved in many ways: one method would be to use different cations, including multivalent ions (e.g., Ca^{2+}, Mg^{2+}). This may not be always possible because of precipitation or other unfavorable interactions with the formation. Another method would be to program the salinity decrease in such a way that the increase in pH is minimized.

A physicochemical model based on the fundamental principles of adsorption and ion exchange in the presence of mass transfer has been developed (Vaidya and Fogler 1990b). As seen from Figures 7 and 8, excellent matches between model predictions and data were obtained. Using a phenomenlogical model (Vaidya and Fogler 1990c) developed recently, we are also able to predict the permeability changes for these cases.

We expect that the results of our study, with minor modifications, will have useful applications in many other areas, such as design of drilling muds, alkaline and surfactant flooding, bacterial transport for enhanced oil recovery, failure of earthen embankments, and most important, groundwater contaminant transport.

High pH and low salinity are detrimental to the permeability of naturally occurring Berea sandstone. Through the mechanism of ion-exchange, salinity shocks (reductions) induce a pH change (increase) and this synergism leads to substantial increase in the double-layer repulsive forces causing drastic reductions in permeability. Finally, a model has been developed, which, in addition to predicting all the experimental observations, resolves conflicts existing in this area for over two decades.

REFERENCES

Khilar, K.C., The water sensitivity of berea sandstone, Ph.D. dissertation, The University of Michigan, Ann Arbor, MI, 1981.

Khilar, K.C. and Fogler, H.S., Colloidally induced fines migration in porous media, *Rev. Chem. Eng.*, 4(1-2):1-149, 1987.

Kia, S.F., Fogler, H.S., and Reed, M.G., Effect of pH on colloidally induced fines migration, *J. Colloid Interface Sci.*, 118(1):158, 1987.

Vaidya, R.N. and Fogler, H.S., Fines migration and formation damage: influence of pH and ion exchange, SPE 19413, Formation Damage Symposium, Lafayette, LA, February 22-23, 1990a.

Vaidya, R.N. and Fogler, H.S., Formation damage due to colloidally induced fines migration, *Colloids and Surfaces*, 50:215, 1990b.

Vaidya, R.N. and Fogler, H.S., An investigation on the permeability reduction in porous media due to fines migration, Paper 275a, American Institute of Chemical Engineers, Chicago, Ill., November 12-16, 1990c.

CHAPTER 21

Effects of Cellular Polysaccharide Production on Cell Transport and Retention in Porous Media

Ray Lappan and H. Scott Fogler

Bacterial reproduction and production of polysaccharides are key factors that segregate bacterial transport in porous media from particle and fines transport. Carefully controlled experiments conducted on both high- and low-permeability ceramic cores have shown that bacteria can plug the pore space and damage the cores. However, further experimentation has demonstrated that polysaccharide production is largely responsible for this damage. In addition, the examination of the axial cell distribution through the damaged cores demonstrated that cell retention by porous medium is higher for cells capable of producing polysaccharides vs those not capable of producing polysaccharides. Injection studies of polymer coated and uncoated cells into a glass micromodel has verified this postulate; cell transport through porous media is strongly affected by their polymeric coating.

For this study, *Leuconostoc mesenteroides* has been selected to be the model bacteria for the transport and retention experiments in porous media because the species has the unique ability to produce dextran, a water insoluble polysaccharide, when fed sucrose. However, when this bacteria is limited to a glucose-fructose feed, no dextran is produced; hence, cellular polysaccharide production can be controlled. This control over cellular production allows for the formulation of experiments that demonstrate the effects of polysaccharide production on cell transport and retention in which polymer production is promoted or repressed.

In addition, *Leuconostoc mesenteroides* bacteria can be considered to be colloidal particles that are self replicating, polymer producing, and are typically, 1 μm in size.

The plugging experiments consisted of two phases: (1) the initial injection of bacteria into the porous medium and (2) the injection of a nutrient feed into the media to promote cell and (depending on the type of saccharide added) polymer production. These experiments were conducted on core samples of both the high and low permeability.

Figures 1 and 2 show the results of plugging of the porous media for both high- and low-permeability core samples. Note that the initial permeability of the high-permeability cores were not affected by the inoculation of cells into the medium. However, the lower-permeability cores did show a two-thirds reduction of permeability, as expected by the relative cell to pore size.

As can be seen from Figures 1 and 2, the permeability of the sucrose-fed cores were drastically affected by polysaccharide production. These results validate the postulate of increased core plugging as the result of polysaccharide production when cell growth rates for sucrose- and glucose-fructose-fed cells are in the same range. A preliminary study of the growth kinetics has shown that this is true. That is, the doubling times for the bacteria are 0.98 h on glucose-fructose feed, and 0.99 to 1 h for the sucrose feed. Equal doubling rates were expected because the nutrient feeds were stoichiometrically equal (i.e., for every mole of sucrose in the polysaccharide-promoting feed there was a mole of glucose and a mole fructose in the polysaccharide repressing feed). Note that sucrose is a disaccharide composed of two monosaccharides, glucose and fructose, joined by a glycosidic bond.

Figure 3 illustrates the staining intensity for the high permeability cores. This figure demonstrates that the cells in the dextrose-fructose fed cores are relatively evenly distributed throughout the core compared with the sucrose fed core, which exhibited a large cell concentration gradient. These results suggest that the polysaccharide-producing cells grew at the injection face, where nutrient availability is high, and were retained there. The nonpolysaccharide-producing cells were not as easily retained by the medium and, thus, could be transported within the core thereby reducing the magnitude of gradient.

To investigate the influence of insoluble dextran on transport properties of *Leuconostoc mesenteroides*, two transport experiments were carried out under analogous conditions with cells grown on a glucose-fructose mixture (Run 1) and cells grown on sucrose (Run 2). During these two experiments, the flow of cells through a glass micromodel was monitored and recorded on videotape. In addition, the inlet pressure and cell concentration at the outlet were recorded. The complete sequence, aimed at examining the reversibility of cell retention involved three steps.

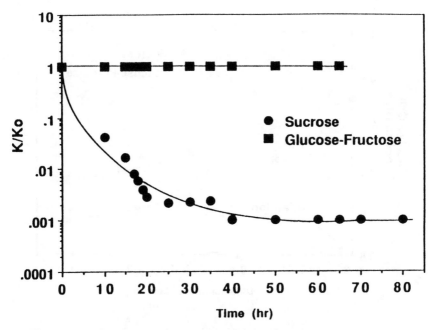

Figure 1. Permeability reduction due to Dextran Synthesis in high permeability core (14.7 Darcies).

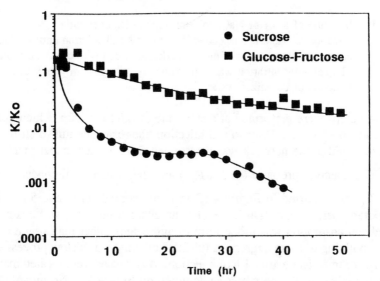

Figure 2. Permeability reduction due to Dextran Synthesis in low permeability core (98 milliDarcies).

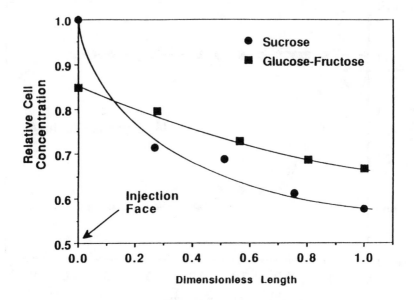

Figure 3. The Axial Cell Concentration (relative to inlet cell concentration of polysaccharide plugged core).

1. Injection of a 0.1 M NaCl solution in the clean network.
2. Injection of bacteria suspended in a 0.1 M NaCl solution until stationery conditions at the outlet of the network (saturation with cells) are reached. The cell concentration was approximately 10^8 cells·cm^{-3} for Run 2.
3. Injection of deionized water.

All steps were performed at a flow rate of 0.02 cm^3/min, which corresponds to ~5 ft/day. Prior to cell injection, the residence time distribution of the fluid in the network was determined from a tracer experiment.

The injection pressure ($\Delta P/\Delta P_0$) and the effluent cell concentration

(C/C_0) are reported in Figure 4 (Run 1: noncoated cells) and 5 (Run 2: dextran-coated cells). During Run 1, no channel plugging was observed, and cell adsorption was in the form of aggregates rather than single cells. The building of these aggregates (clusters of cells) was a slow process that only became visible after 1 h of injection. Aggregates were located mainly on the edges of the network, as depicted on Figure 6a. No meaningful variation in pressure was recorded during the run, and the shape of the breakthrough curve showed a temporary decrease when the feed was shifted to deionized water (see Figures 2 and 3). This decrease in pressure is possibly the result of switching the feeds from concentrated cell suspension

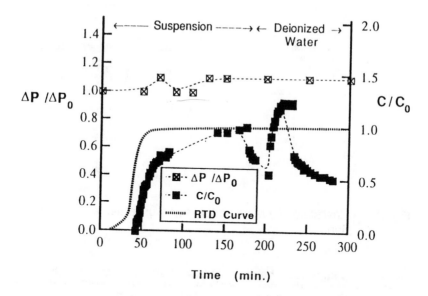

Figure 4. Normalized pressure drop, cell's breakthrough curve and RTD in the course of run 1.

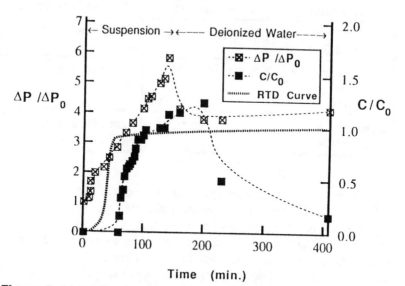

Figure 5. Normalized pressure drop, cell's breakthrough curve and RTD in the course of run 2.

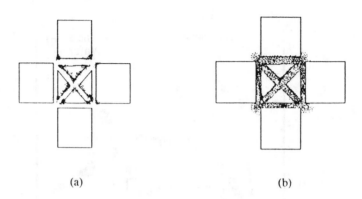

(a)   (b)

Figure 6. Final state of a representative network's element: (a) Run 1, (b) Run 2.

Table 1.   Comparison between the two transport experiments.

Characteristics	Run 1 (non-coated cells)	Run 2 (dextran-coated cells)
Pressure Drop	Almost Negligible	$\Delta P(Final)/\Delta P(Initial)=5.6$
Localization of Absorbed Cells	Network's Edges	All the Available Surfaces
Cell Absorption at Saturation (% of injected cells)	$20(7.2 \ 10^7 \ cells)$	$40(2.2 \ 10^9 \ cells)$
Desorption and Remediation	Partial Desorption	Partial Desorption No Remediation

to deionized water causing readjustments of pressure within the syringes.

During Run 2 (see Figure 5) channel plugging was observed after 2 h of injection. Adsorbed cells were in the form of aggregates rather than single cells. Examination of the video recording showed that the building up of aggregates on the walls was faster and more significant than that seen in Run 1. In addition, aggregates invaded the void space, as well as all the surfaces. A rapid and dramatic increase in pressure was recorded during the adsorption step, as well as during the first stage of deionized water injection. At the end of this step, the pressure decreased partially to reach about four times the initial pressure.

Comparison between Run 1 and 2 shows that the presence of dextran around the cells dramatically reduces cell transport efficiency. Thus, cell—surface interaction resulting from the presence of polymer is one of the key mechanisms in the transport of bacteria through porous media. This specific interaction has to be taken into account for the development of any transport model.

This effect may be seen as beneficial for profile modification. Indeed, in a first step noncoated cells can be transported efficiently across the reservoir and, once adsorbed, may reduce local heterogeneities, provided that in a second step sucrose is injected in the formation. Main comparisons between Runs 1 and 2 are reported in Table 1.

REFERENCE

Lappan, R.E. and Fogler, H.S., The effects of bacterial polysaccharide production on formation damage, SPE 19418, Formation Damage Symposium, Lafayette, LA, February 22-23, 1990.

SUMMARY AND RESULTS OF

SECTION III

MANIPULATING COLLOIDS TO CHANGE AQUIFER PERMEABILITY

CHAPTER 22

Discussion Leaders

H. Scott Fogler and James R. Hunt

This group identified a number of manipulative strategies for altering the permeability of aquifers. The strategies are outlined on Table 22.1 and each strategy is described more fully below.

APPROACH 1: STRAINING OF PARTICLES IN PORE RESTRICTIONS

Description

Filtration theory distinguishes different mechanisms for the removal of suspended particles by porous media. The process considered under this heading is limited to removal of particles and colloids by physically capturing particles in pore restrictions that are smaller than particles.

Processes

This is a purely physical process of retaining solids by porous and fibrous materials because the particle size exceeds the pore size.

Advantages

There is considerable industrial experience in cake filtration, and in the design of geotechnical filters for the stabilization of soil. No chemistry is involved in retaining the particles, and very high permeability reductions are possible. Predictive network models are available for two dimensional systems.

Limitations

Straining is a very localized phenomenon and very little penetration is expected. A well would not be a good delivery system, and since straining is very dependent on the overlap of particle and pore sizes, aquifer hetero-geneities would have to be known at the pore level.

Gaps in Knowledge

Predictive models are dependent on knowing the pore size distribution, which is not measurable in natural systems. Models are unavailable for three dimensional networks due to the computational burden. Thus far, there is no known way to scale laboratory results to field conditions.

Overall Feasibility

Straining is a very localized process that can greatly reduce aquifer permeability over distances on the order of centimeters which severely limits utilization of this method to reduce permeability in an aquifer.

APPROACH 2: DEEP-BED FILTRATION OF GROUNDWATER COLLOIDS

Description

Water filters operate by removing particles much smaller than the pore size of a filter and accumulate the particles within the filter over distances of less than one meter. Water velocities are 100 to 400 m d^{-1} in deep bed filters, and the filters require frequent backwashing to remove the accumulated solids and restore the filter's permeability. Particle capture onto fixed surfaces is promoted by the addition of chemical coagulants that destabilize particles so they will aggregate with themselves and efficiently stick to solid surfaces.

Processes

Several processes control the magnitude and effectiveness of deep-bed filtration, including:

- *Aggregation.* Deep-bed filtration requires the collision and sticking of small particles onto filter surfaces. The usual practice is to add coagulants to increase sticking efficiency, but particle aggregation and particle filtration do occur under natural water conditions, but at reduced rates.

- *Restabilization.* For particles retained on fixed surfaces and particle aggregates held together by weak attractive forces, changes in solution composition can restabilize the particles and cause them to be resuspended. Low ionic strength solutions will cause restabilization as will changes in pH.

- *Shear.* As particle aggregates accumulate within porous media the fluid shear imposed on the aggregates can increase and cause resuspension or erosion of the aggregates back into the moving fluid. Resuspended aggregates are greater in size than the primary particles initially present in suspension.

Advantages

There is considerable experience in deep bed filtration for dilute suspensions as encountered in water treatment. This experience is for one dimensional flow in homogeneous media that is only 1 to 2 meters thick. Engineers have been able to optimize particle removal during filtration by selection of loading rates, media sizes, and chemically conditioning the particles. Considerable research has been undertaken on single particle collection by clean media surfaces.

Limitations

Models developed for particle removal and permeability reduction during the course of filtration are empirical and cannot be extrapolated to natural systems. The processes of deposition and erosion by shear are stochastic and not easily modeled on a microscale.

Gaps in Knowledge

Application of deep-bed filtration concepts to natural aquifer materials and groundwater chemistries requires predictive models of aggregate structure within porous media that accounts for permeability reduction and flow redistribution.

Overall Feasibility

The empirical data and theories are currently insufficient in reliably manipulating aquifer permeability at the field scale.

APPROACH 3: "LOG-JAMMING" PARTICLES AT PORE RESTRICTIONS

Description

Particles introduced into porous media or resuspended from porous media can accumulate at pore restrictions and substantially reduce permeability. The effect requires particle sizes comparable to the minimum pore opening and high particle concentrations.

Processes

The principal process considered here is blockage of pores as particles are physically trapped when too many attempt to flow through a constriction.

Advantages

This process can be initiated deep within a formation by changes in solution composition and flow direction to resuspend a high concentration of previously deposited particles.

Limitations

There is no theory for this process and only empirical observations in sandstone cores where clays are not much smaller than the pore restrictions. The process would probably not have application to more permeable aquifer materials.

Gaps in Knowledge

Predictive models of erosion rates as functions of fluid shear and solution chemistry are needed to extrapolate laboratory observations on sandstone cores to other conditions.

Overall Feasibility

For sandstones of importance in petroleum reservoirs, permeability manipulation is feasible. However, there is no experimental or theoretical experience in more permeable, unconsolidated porous media.

APPROACH 4: INDUCED IN SITU PRECIPITATION OF PARTICLES

Description

If a mineral phase could be precipitated in situ, then significant permeability reductions could occur. One compound that is a possibility is $BaSO_4$, if barium and sulfate could be mixed in the formation in the correct stoichiometric ratio.

Processes

Two key processes must be understood and controlled in situ to accomplish this strategy:

- *Precipitation kinetics.* The clogging of the formation requires the precipitation of a solid within the formation at a rate fast enough to block the pore space, and forming a solid matrix sufficiently strong to resist hydrodynamic shearing.

- *Mixing of reactants in situ.* To achieve precipitation of an inorganic solid within a permeable formation requires the delivery of the reactants in stoichiometric amounts over a wide three-dimensional region and then have the reaction proceed to produce pore plugging solids.

Advantages

This is an in situ process that would result in a permanent reduction in permeability if further fluid exchange is prevented. With no flow, the solution chemistry would remain in equilibrium with the solid precipitate.

Limitations

This technique has not been tried in the subsurface and would be difficult to engineer because of the requirements for mixing in heterogeneous media. There will always be a concern about introducing a foreign substance into the subsurface environment that may migrate under unexpected conditions.

Gaps in Knowledge

There does not exist field or laboratory research on chemical kinetics nor microscale mixing in heterogeneous media to predict performance at this time. The structure and strength of the precipitate will be as important for this strategy as in deep-bed filtration and biofouling.

Overall Feasibility

The feasibility of this strategy for aquifer permeability reduction is unknown at this time.

APPROACH 5: BIOFOULING-INDUCED PERMEABILITY REDUCTIONS

Description

Microorganisms, when given more substrate than required for maintenance, will produce new cells and extracellular polymers that will occupy space within porous media. Microorganisms can cause permeability reductions to less than 1% of the original permeability in laboratory column experiments and biofouling of well screens has been observed when substrates are injected into the subsurface.

Processes

Application of biofouling to issues of aquifer permeability is quite complex and requires consideration of several biological and physical processes, including:

- *Cell growth.* When new cells are produced they decrease the pore space available to fluid flow and thus decrease permeability. Some of the new cells may slough off into the pore fluid and be carried down gradient, but most cells in aquifer materials are found associated with surfaces.

- *Extracellular polymer production.* Microorganism, besides producing more cells, also secrete extracellular polymers or slimes that attach to cells and media surfaces and alter fluid flow through a local reduction in permeability.

- *Anaerobic gas production.* Under methanogenic and denitrification conditions microorganisms can produce methane and nitrogen gas at concentrations that exceed their solubility in water. Gas bubble formation in the pore spaces can cause a significant reduction in water permeability due to relative permeability effects.

- *Substrate supply.* Microorganisms have requirements for multiple substrates, trace metals, and occasionally vitamins. All of these substrates must be available at the scale of the microorganism at the correct stoichiometric ratios to achieve growth.

- *Predation.* Besides bacteria that convert organic and inorganic materials into energy and cells, there are organisms that control bacterial populations and limit the longevity of any biofilter or bioplug.

Advantages

Microorganisms are ubiquitous in the subsurface environment and mixed cultures can be stimulated to consume many common organic compounds to produce more cells and extracellular polymers. Mixed cultures are very efficient in optimizing growth under the conditions available and empirical evidence shows that permeability reductions to less than 1% of the original permeability are easily achieved over distances on the order of ten centimeters.

Limitations

As with other chemical reactions, all the reactants must be present at each cell for the reactions to proceed. This strategy is highly dependent on mixing at the pore scale and results thus far indicate that little penetration into a formation is likely when substrates and nutrients are injected together. Pulsation of various nutrients over time at injections wells has caused mixing in the formation and has achieved more dispersed growth. There is concern on the ability to deliver enough substrate to clog a formation and on the duration of such a bioplug.

Gaps in Knowledge

Current research on implementing bioremediation in aquifers is attempting to avoid biofouling near the injection wells. The process therefore occurs in situ; however, achieving a sufficient degree of biofouling in a heterogeneous formation sufficient to reduce regional flow through a contaminated zone has not been demonstrated. Again, microscale mixing in aquifers combined with a better understanding of the aquifer heterogeneity are essential in implementing this strategy in the field. There are instances where a particular bacterium may have preferred biofouling capabilities, in which case concerns of organism delivery and survival with indigenous organisms must be addressed. For bacteria that are 1 μm in size, organism seeding and growth in low permeability regions less than 0.1 darcy could be a problem. The microorganisms would be excluded from those regions by their size.

Overall Feasibility

Biofouling does occur when stoichiometric amounts of substrates are present at pores containing adapted microorganisms. Biofouling has usually been a hinderance to bioremediation of aquifer contamination and experience is not available to evaluate its suitability for aquifer plugging with minimal fluid injection and maximum permeability reduction.

APPROACH 6: MULTIPHASE FLOW OF IMMISCIBLE FLUIDS IN POROUS MEDIA

Description

Because of interfacial tension between two immiscible fluids, there exists a reduction in each fluid's permeability when both fluids are flowing through the porous media. The permeability to water flow can be reduced to less than 1% of the original permeability as the other fluid becomes the dominant phase. Displacement of one phase by another can erode attached colloids because of the interfacial tension between the fluids and the interaction of these fluids with the mineral surface.

Processes

Two distinct processes related to interactions of immiscible fluids can effect permeability reductions:

- *Permeability reduction.* Multiphase flow through porous media causes a reduction in each fluid's permeability because of interfacial tension.

- *Wettability and particle erosion.* Mineral surfaces interact with fluids at their surfaces. Most mineral surfaces are water-wetted such that organic solvents and petroleum products would be excluded from the interface. As a nonwetting fluid displaces a wetting fluid, the colloids and particles attached to minerals surfaces can be eroded by the other liquid. The displaced particles can jam downstream pore throats and further reduce the permeability for both fluids.

Advantages

Displacement of water from a porous media by a nonaqueous phase liquid can reduce the formation's permeability two ways. First, by relative permeability effects, and second, by eroding previously deposited particles and having them removed by straining at pore restrictions.

Limitations

The major limitation of this strategy is the introduction of a nonaqueous phase liquid into a subsurface environment. Nonaqueous phase liquids are most likely contaminants in their own right and any attempt to inject them into a contaminated area would only meet with scientific and regulatory opposition.

Gaps in Knowledge

Predictive models are under development for multiphase flow in porous media. The use of multiphase flow to erode particles or to block media pores with eroded particles is not sufficiently developed to predict the field application of this strategy. Furthermore the feasibility of this mechanism in higher permeability aquifers has not been demonstrated.

Overall Feasibility

Multiphase flow through porous media could mobilize trapped colloids and could reduce permeability to water flow by particle blockage of pores and relative permeability effects. The feasibility of implementing this strategy in the field is not known.

DISCUSSIONS AND CONCLUSIONS

The group agreed that there were several conceptual strategies that offered long-term potential for manipulating aquifer permeability by controlling colloid production and deposition. Changing the aqueous chemistry to restabilize previously deposited colloids or increasing fluid flow can erode colloids and increase formation permeability. Alternatively, flow and solution conditions can reconfigure deposited colloids into configurations that result in significant reductions in formation permeability. Introduction of particles or substrates for microorganism growth will result in permeability reduction by particle accumulation or biofilm development.

Of the several options considered, biofouling appeared to excite the greatest interest among the participants because it has been observed in the field. Although the potentially temporary nature of the permeability reduction may be a problem in some applications, this strategy seemed particularly well-suited to enhancing in situ treatment (e.g., soil-washing or air-stripping) of small areas. In this scenario, biofouling could be used to temporarily "wall-off" the zone to be treated and help confine the treatment to the target area.

However, it must be cautioned that experience on permeability alteration is largely confined to laboratory systems which have been driven by experience in petroleum reservoirs. The dominant mechanisms of particle attachment, erosion, and biofilm development have been identified, but predictive models are not available that apply to natural or degraded subsurface environments.

TABLE 22.1 Evaluation of strategies considered in group discussions.

PROPOSED MANIPULATION APPROACH	KEY HYDRO-GEO-BIO-CHEMICAL PROCESSES IN APPROACH	ADVANTAGES/ REASONABLENESS OF APPROACH	LIMITATIONS/ IMPRACTICALITIES OF APPROACH	MAJOR GAPS IN KNOWLEDGE	OVERALL FEASIBILITY
1. Straining of particles in pore restrictions	Physical	Industrial experience in cake filtration No chemistry involved Predictive 2-D models network models available	Very localized phenomena Little penetration expected Aquifer heterogeneities must be known at pore level	Pore and particle distributions must be known 3-D network models unavailable Scaling from lab to field	Limited due to very localized nature of process
2. Deep-bed filtration of groundwater colloids	Aggregation Restabilization Shear	Experience in homogeneous media for water treatment	Models empirical Processes are stochastic	Predicting deposit permeability and flow redistribution	Limited due to insufficient empirical data or theory at field scale
3. "Log-jamming" particles at pore restrictions	Blockage of pore openings	Potential for deep penetration	No theory Experience limited to sandstone cores	Predictive models to generalize understanding	Feasible in sandstones Limited for more permeable, unconsolidated media
4. Induced in situ precipitation of particles	Precipitation kinetics Mixing of reactants in situ	In situ production of reduced permeability zone	Untried Difficulties in mixing within heterogeneous media	Predictive capability	Unknown due to lack of experience

PROPOSED MANIPULATION APPROACH	KEY HYDRO-GEO-CHEMICAL PROCESSES IN APPROACH	ADVANTAGES/ REASONABLENESS OF APPROACH	LIMITATIONS/ IMPRACTICALITIES OF APPROACH	MAJOR GAPS IN KNOWLEDGE	OVERALL FEASIBILITY
5. Biofouling-induced permeability reductions	Cell growth Extracellular polymer production N_2 and CH_4 gas production Nutrient supply Predation	Biofouling does occur in aquifers Permeability reduction to less than 1%	Mixing at the pore scale Limited penetration into formation Transitory effect on permeability?	Capabilities of indigenous micro-organisms Microscale mixing Size exclusion may limit organism seeding and growth	Promising since biofouling and permeability reductions have been document-ed.
6. Multiphase flow of immiscible fluids in porous media	Interfacial tension Relative permeability Particle erosion	Water permeability reduction by two possible processes	Regulatory restrictions on introduction of nonaqueous fluids into aquifers	Predictability Feasibility, especially in higher permeability aquifers	Not known

SECTION IV

MODIFIED COLLOIDS, SURFACTANTS, AND EMULSIONS

CHAPTER 23

Manipulating Colloids and Contaminants in the Subsurface Via Surfactants

Richard G. Luthy

MODERATOR'S OVERVIEW

It is clear that there are many possible potentially useful applications of surfactants and emulsions for in situ subsurface remediation, but there are major gaps in understanding coupled chemical and physical processes that affect phase partitioning and transport of surfactants, colloids, and contaminants. There are many complex interactions affecting the utility of such applications because surfactants may mobilize or immobilize inorganic/organic compounds, microorganisms, and colloids.

The role that surfactants may play in modifying or immobilizing colloids and/or contaminants in the subsurface may be considered from several perspectives. Some opportunities for beneficial use of surfactants to manipulate the behavior of organic compounds and colloids in the subsurface include the following.

- Surfactants can alter sorption and partitioning reactions of hydrophobic organic compounds (HOCs). This may be beneficial for remediation in certain situations, and the physics and chemistry of surfactant-aided dissolution of sorbed organic contaminants via micellization are being explored currently in several laboratory studies with heterogeneous materials.
- In another approach, surfactants may alter the wetting characteristics of nonaqueous phase liquids (NAPLs), and thus assist recovery of NAPLs through solubilization or displacement.
- Clays may be intercalated with cationic surfactants, and this may be beneficial for development of manipulated colloids with high sorptive properties for HOCs.

- The rate and extent of bioremediation of organic compounds in soils or aquifer materials may be limited by the desorption or solubilization of organic compounds. Surfactants, synthetic or microbially produced, may assist the bioremediation process by lowering interfacial tension and/or solubilizing organic compounds and facilitating availability to microorganisms.

Surfactants may be used beneficially for remediation of sorbed organic contaminants, and for remediation of dense nonaqueous phase liquids (DNAPLs). These are different, but related, problems. A "detergency"-based, surfactant-enhanced remediation approach would employ surfactant to solubilize sorbed HOCs. This approach might be used for in situ soil washing, or possibly to accelerate bioremediation by increasing the accessibly of HOCs to a microorganism. An "oil recovery"-based, surfactant-enhanced remediation approach would employ surfactant to cleanup DNAPL in porous media. In this strategy surfactant is used to decrease the interfacial tension between DNAPL and water in order to recover the organic liquid by miscible displacement.

The engineering-science literature on surfactant-enhanced soil remediation is relatively barren. Although inferences can be drawn from the fields of detergency science and enhanced oil recovery, the application of these techniques to groundwater systems is largely unexplored with regard to elucidating key problems and fundamental concepts. There are examples of bench-scale tests, and some pilot-scale work, but these studies have been principally trial-and-error, phenomenological investigations.

The behavior of surfactants in soil-and-water systems is very complex. Current research is only beginning to understand surfactant interaction with soil media, and the role of such interactions on wetting, micellization, surfactant sorption, and surfactant degradation. Anionic surfactant behavior is especially sensitive to ionic composition, since these surfactants may interact with calcium and magnesium and precipitate from solution. Nonionic surfactants sorb onto porous media. These type of processes and other reactions affecting surfactant micellization and phase behavior are especially important in remediation of HOCs and DNAPLs. At present, there is a lack of fundamental predictive techniques for rational design of surfactant-enhanced remediation technologies. This problem becomes particularly complicated in consideration of the numerous types of surfactants commercially available, which may be used singularly or in combination.

For these various reasons, it is necessary to have knowledge of the physicochemical behavior and environmental fate of surfactants in soil and groundwater systems in order to effectively design and implement field tests with surfactants. Laboratory tests can help define mechanisms governing the interaction among surfactants, soils, and hydrophobic organic compounds. A clear understanding of such microscale phenomena will provide a better picture of macroscale effects.

  i.   What properties of surfactants govern sorption onto subsurface media? How may this be predicted?
  ii.   What are the rates of desorption of surfactants? Can a water flush recover sorbed surfactants?
  iii. How may water/surfactant/HOC mixtures be treated for surfactant recovery and reuse?
  iv.   What are the long term consequences of surfactant residuals in the subsurface?

Some tests have been conducted to assess the effect of surfactant-enhanced HOC solubility on microbial mineralization. This work is not extensive, and has given somewhat varied results. Clearly much work needs to be done to understand how surfactants may influence bioremediation. Some questions relative to this work are:

  i.   How readily are surfactants biodegraded? Does sorbed-phase surfactant degrade at a rate different than micellar-phase surfactant?
  ii.   How does surfactant structure affect performance as a biodegradation catalyst? What are the relationships between activity and surfactant concentration in soil and aqueous phase?
  iii. How do surfactant micelles affect microbial cell membrane structure and transport of HOC? What mechanisms may result in interference of either the transport mechanism for HOC across the cell membrane, or of the activity of membrane-bound protein.

It would appear that the most rapid progress on surfactant-enhanced bioremediation may be achieved by parallel research investigations on HOC mineralization and physicochemical partitioning phenomena, in conjunction with study of the interaction of surfactants and cell membranes.

Surfactant solutions may be beneficial in assisting recovery of DNAPLs from the subsurface. The key mechanisms for miscible displacement are high interfacial tension lowering (i.e., dramatic increase of the capillary number) and the solubilizing capacity of the surfactant solution. The solubilizing capacity of a surfactant is best described by a ternary phase diagram of surfactant/water/organic systems. While such diagrams have been generated for enhanced oil recovery applications, there is not a corresponding body of literature on those systems that may be used for environmental restoration.

Some of the key considerations in surfactant-enhanced miscible displacement are:

i. What candidate surfactants are environmentally acceptable? May relatively nontoxic surfactants be employed successfully for miscible displacement of DNAPLs?

ii. Which potentially useful surfactants may sorb onto soil and thus lose their ability to displace DNAPL?

iii. What are the ternary phase relationships for surfactant/water/organic systems?

Surfactants may be used to manipulate the behavior of colloids in the subsurface. The mobility of colloids may be increased or decreased through electrostatic/chemical modification of the surface of the colloid or porous media matrix. Surfactants may also influence the rate of nucleation and growth of colloidal particles, which subsequently serves to influence the mobility of sorbable pollutants. Many inter-related mechanisms may lead to formation of colloids and affect colloid stability, with each mechanism responding in a different way to surfactant addition.

The most promising approaches for surfactant-aided remediation of subsurface systems appear to be: (i) mobilization of pollutants with surfactants, and (ii) use of surfactants to alter aquifer permeability.

REFERENCES

Aronstein, B.N., Calvillo, Y.M., and Alexander, M., Effects of Surfactants at Low Concentrations on the Desorption and Biodegradation of Sorbed Aromatic Compounds in Soils, Department of Soil, Crop and Atmospheric Sciences, Cornell University, Ithaca, New York, 1991.

Edwards, D.A., Luthy, R.G., and Liu, Z., Solubilization of Polycyclic Aromatic Hydrocarbons in Micellar Nonionic Surfactant Solutions, *Environ. Sci. Technol.*, 25(2):127-33, 1991.

Laha, S. and Luthy, R.G., Mineralization of Phenanthrene in Soil-Water Systems with Nonionic Surfactants, *Environ. Sci. Technol.*, (in press), 1990.

Liu, Z., Laha, S., and Luthy, R.G., Surfactant Solubilization of Polycyclic Aromatic Hydrocarbon Compounds in Soil-Water Suspensions, *Water Sci. Technol.*, 23:475-85, 1991.

CHAPTER 24

Effects of Surfactants on NAPL Mobility in the Subsurface

Kim F. Hayes and Avery H. Demond

In recent years, contamination of groundwater by organic liquids has become widespread in the United States. Because many organic liquids have a low solubility in water, they form a separate organic liquid phase or a nonaqueous phase liquid (NAPL) in groundwater. The most common remediation process of an aquifer where a NAPL has been identified involves the extraction of the mobile organic liquid and the extraction and treatment of the surrounding groundwater (Williams and Wilder 1971; EPRI and EEI 1988). However, such a remediation process may leave behind as much as 30% of the NAPL entrapped in pore spaces (Thornton 1980). This residual quantity may then dissolve slowly into the surrounding groundwater, thus serving as a long-term source of contamination (Hunt et al. 1988). Consequently, it is important to examine cleanup techniques that may reduce this residual and, in turn, reduce the potential for continued contamination of the aquifer.

One remediation technique that is currently under consideration is surfactant flushing. Surfactant flushing may enhance the mobility of NAPLs through a variety of mechanisms. For example, when surfactants are injected into the subsurface, they can lower the liquid-liquid interfacial tension or increase the aqueous phase solubility. Either of these effects would be expected to enhance NAPL mobility. In addition, surfactants may affect the wettability and NAPL mobility of the system by sorbing and changing the interfacial tension at the three relevant interfaces: the organic liquid-water, solid-water, and solid-organic liquid interfaces. Such effects need to considered in designing strategies for aquifer remediation by surfactants. In this paper, the mechanisms by which surfactants affect NAPL mobility are briefly reviewed, followed by a discussion of recent work illustrating how changes in wettability by surfactants affect NAPL transport properties.

When surfactants are added to water-NAPL systems, significant changes in interfacial properties will result that may lead to dramatic changes in NAPL mobility. Surfactants may affect NAPL mobility by (1) lowering the water-organic liquid interfacial tension, (2) increasing the organic compound's aqueous solubility, or (3) changing the wettability of the system. The changes in NAPL mobility that result from the addition of surfactants result from the unique physicochemical properties of surfactants. Surfactants are amphipathic molecules, possessing both a hydrophobic, alkyl hydrocarbon chain end (referred to as the "tail") and a polar, hydrophilic functional group portion (referred to as the "head"). Because of this dual hydrophobic-hydrophilic character, surfactant molecules usually concentrate at interfaces. At an organic liquid-water interface, surfactant molecules will align themselves with the hydrophobic tail in the organic phase and the polar head group in the water phase. This interface structure results in a lower interfacial tension (IFT) between the two liquids. IFT lowering is one method that can be used to facilitate the removal of entrapped or residual NAPLs. As the IFT approaches zero, the pressure necessary to displace the organic liquid approaches zero, and as a result, the organic liquid may be removed through immiscible displacement by pumping water through the aquifer. Yet, previous studies in the oil industry suggest that the IFT needs to be reduced to considerably less than 1.0 dyne/cm before sizable reductions in residual saturation are achieved (Foster 1973; Wagner and Leach 1966; Ling et al. 1987). Consequently, approaches emphasizing increased solubilization may be more efficient.

To achieve maximum solubilization, the critical micelle concentration of the surfactant must be exceeded (Rosen 1989). When the surfactant concentration in water becomes sufficiently high, unfavorable entropic effects between the water molecules and the hydrophobic-tail portion of the surfactant lead to the formation of micelles or dispersed aggregates of surfactant monomers. In water, the micelles consist of a hydrophobic inner-core made up of hydrocarbon tails and a hydrophilic outer surface made up of polar head groups. In the presence of NAPLs, surfactants may also lead to the formation of NAPL-in-water emulsions, which are stable dispersions of organic-liquid droplets in the water phase. The effects of emulsions on NAPL mobility depend on the emulsion size and phase behavior. When micelles or microemulsions are formed, the rheological properties of the organic-in-water emulsion remain essentially unchanged from the initial water phase. As a result, microemulsion dispersions containing high concentrations of NAPLs may be removed with the water phase, thus enhancing NAPL cleanup. On the other hand, when macroemulsions are formed, the resulting NAPL-in-water emulsion may have a significantly increased viscosity compared with pure water, which can lead to pore clogging. The relative efficacy of a given surfactant for forming microemulsions or macroemulsions and enhancing NAPL solubility depends on the phase behavior of the NAPL-water-surfactant system. Although significant progress has been made in enhanced oil recovery research in understanding phase behavior of oil-water-surfactant systems, much remains to be done in groundwater systems.

Because of high surface activity, surfactants added to promote changes in IFT and solubilization may also alter the wettability of the system. Wettability qualitatively describes the degree to which a liquid spreads on a solid. It is often defined on the basis of the contact angle, the angle formed at the junction of the three phases, which is based on the force balance of the three interfacial tensions of the system (Hiemenz 1986)

$$\cos \theta = \frac{\gamma_{L_1 S} - \gamma_{L_2 S}}{\gamma_{L_1 L_2}} \tag{1}$$

If the contact angle through the reference liquid (here designated as the droplet liquid) is $< 90°$, then the reference liquid wets the solid; conversely, if the angle formed is $> 90°$, then the other liquid wets the solid. If the wetting liquid is water, then the surface is referred to as hydrophilic; in contrast, if the wetting liquid is an organic, then the surface is termed hydrophobic.

The wettability of a particular system may vary from its native state of hydrophilicity to a state of hydrophobicity as a function of the surfactant concentration sorbed. Preliminary results from current work in our laboratories show that it is possible to convert a silica surface from hydrophilic to hydrophobic and then back to hydrophilic as a function of increasing CTAB (cetyltrimethylammonium bromide) surface coverage. This wettability conversion can be explained based on an analysis of the configuration of sorbed CTAB. A direct measure of the configuration of sorbed CTAB molecules would require surface spectroscopy. However, in absence of direct structural information, the configuration of CTAB molecules can be inferred from sorption isotherm and electrophoretic mobility data (Figures 1 and 2). For example, the CTAB sorption isotherm may be divided into four regions (Figures 2 and 3). In region I, isolated CTAB molecules are thought to be sorbed by an electrostatic attraction between negatively charged surface sites and the positively charged surfactant. In this region, the silica is hydrophilic because the surface is still composed of mostly deprotonated surface hydroxyl groups, as indicated by the negative value of the electrophoretic mobility, a surrogate measure of the interfacial potential. The marked increase in sorption in region II is thought to result mainly from hemimicelle formation, a result of lateral interaction between the hydrocarbon chains. The electrophoretic mobility is negative in the absence of CTAB, becomes less negative with the addition of CTAB, and finally changes sign. As the CTAB concentration increases

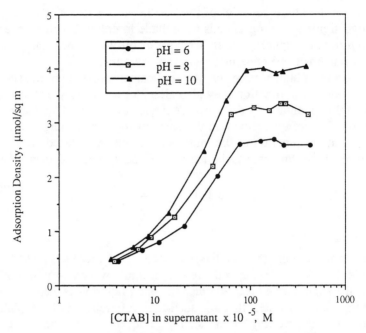

Figure 1. Adsorption Isotherm for CTAB on Silica (0.5 to 10μ size fraction, 50g/l, background electrolyte 0.01N NaCl).

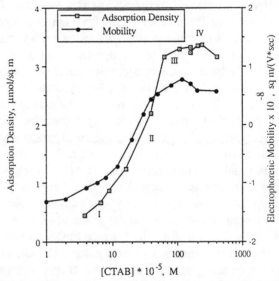

Figure 2. Adsorption Density and Electrophoretic Mobility for CTAB on Silica (0.5 to 10μ size fraction, 50g/l, background electrolyte 0.01N NaCl, pH = 8).

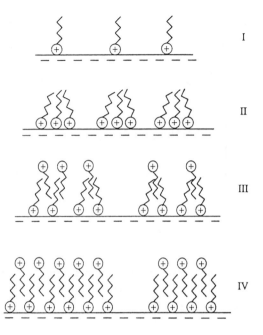

Figure 3. CTAB Sorption as a function of surface coverage: Region I, electrostatic sorption; Region II, onset of hemimicelle formation; Region III, hemimicelle formation and onset of bilayer sorption; Region IV, complete bilayer formation.

further, the slope in the isotherm decreases (region III), reflecting a lower affinity of a positively charged surfactant for a positive surface. The increase in sorption in this region is thought to result from the favorable hydrophobic interaction between the long hydrocarbon portions of the surfactant molecules. This strong hydrophobic interaction may also lead to bilayer formation, which is thought to be negligible in region II but substantial in region III. Because the interface is composed of mostly hydrocarbon tails, a more hydrophobic surface results in regions II and III. Finally, the sorption isotherm reaches a plateau (region IV) near the critical micelle concentration (CMC) of CTAB ($9.0 \times 10^{-4}M$) when a complete bilayer is formed on the surface. The sorption density remains constant at this point because the formation of micelles is energetically more favorable than additional sorption. Because the water-exposed portion of the interface is made up mostly of positively charged head groups in region IV, the surface becomes more hydrophilic and water-wet again in this region.

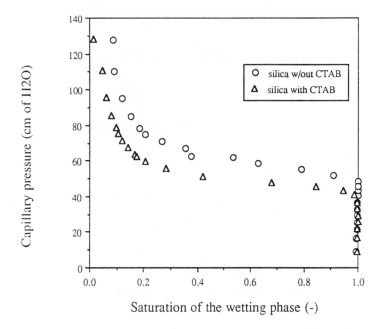

Figure 4. Capillary pressure-saturation relationships for air-water in CTAB-coated and uncoated silica systems.

The state of wettability plays a central role in the phenomenon of two-phase flow in porous media. When two immiscible phases coexist in a porous medium, they are separated by a series of menisci, the positions of which effectively determine the networks through which the two phases flow. In general, the wetting liquid tends to occupy the smaller pores,whereas the nonwetting liquid occupies the larger ones. Because the menisci positions are a function of interfacial forces, these forces are significant in defining the transport pathways of the phases. In the modeling of two-phase flow, these effects show up in two constitutive relationships, relative permeability-saturation and capillary pressure-saturation.

Work by Demond and Roberts (1991a,b) shows that the capillary pressure-saturation relationship is more sensitive than the relative permeability-saturation relationship to small changes in the contact angle outside of the vicinity of 90°. Thus, the effect of wettability on two-phase flow is perhaps best indicated by changes in capillary pressure.

According to the Young-Laplace equation, written here for a spherical meniscus in a capillary tube of radius r, the capillary pressure is given by

$$P_c = \frac{\gamma_{L_1 L_2} \cos\theta}{r} \qquad (2)$$

where P_c, the capillary pressure, is the difference in pressure between the nonwetting liquid and wetting liquid, is the interfacial tension between the two liquid phases, and θ is the contact angle. This equation suggests that as a system becomes more hydrophobic, the pressure necessary to displace one phase from a pore of a given radius decreases.

The effects of changing the wettability on the capillary pressure-saturation relationship is shown in Figure 4. The first system is composed of silica without CTAB sorbed to the surface, water, and air. It shows a displacement pressure, or the capillary pressure at which the first significant drainage occurs, of about 50 cm of water. The second system is composed of silica to which CTAB has been sorbed at a surface coverage of about 7 mol/m^2, water and air. This system shows a displacement pressure of about 40 cm of water. In both systems the liquid-air interfacial tension is nearly the same. Thus, the 20% decrease in displacement pressure is attributable to the changes in the contact angle resulting from the sorption of surfactant at the solid surfaces. Currently, work is being conducted to quantify the effects of changes in wettability caused by CTAB sorption onto silica on the capillary pressure-saturation relationship of xylene-water systems.

Surfactants and interfacial properties play an important role in controlling NAPL mobility in subsurface systems. Ultimately, it would be desirable to develop a quantitative theory for predicting the changes in the important multiphase flow constitutive relationships from changes in interfacial properties. Through the combined measurements of macroscale transport properties and microscale interfacial properties such as those discussed here, it is hoped that an improved method for evaluating transport relationships based on the physical-chemical characteristics of a system will result.

REFERENCES

Demond, A.H. and Roberts, P.V., Effect of interfacial forces on two-phase capillary pressure relationships, *Water Resour. Res.*, 27(3):423-37, 1991a.

Demon, A.H. and Roberts, P.V., Estimation of two-phase relative permeability relationships for organic liquid contaminants, (submitted), 1991b.

Desai, F.N., Demond, A.H., and Hayes, K.F., The influence of surfactants sorption on capillary pressure-saturation relationships, in ACS Symposium volume from Colloid and Interfacial Aspects of Groundwater and Soil Cleanup, 65th

Annual Colloid and Surface Science Symposium, American Chemical Society, Norman, OK, June 17-19, 1991, 1992.

Electrical Power Research Institute (EPRI) and Edison Electric Institute (EEI), 1988, Remedial Technologies for Leaking Underground Storage Tanks, Lewis Publishers, Chelsea, Mich.

Foster, W.R., A low-tension waterflooding process, *Petroleum Technol.*, 25:205-10, 1973.

Hunt, J.R., Sitar, N., and Udell, K.S., Nonaqueous phase liquid transport and cleanup, 1. Analysis of mechanisms, *Water Resour Res.*, 24:1247-58, 1988.

Hiemenz, P.C., *Principles of Colloidal and Surface Chemistry*, Marcel Dekker, Inc., New York, 1986.

Ling, T.F., Lee, H.F., and Shah, D.O., Surfactants in enhanced oil recovery, in *Industrial Applications of Surfactants*, Karsa, D.R., Ed, Royal Society of Chemistry, London, 1987, 126-78.

Rosen, M.J., *Surfactants and Interfacial Phenomena, 2nd ed.*, John Wiley and Sons, 1989.

Thornton, J.S., Paper presented at the National Conference on Control of Hazardous Material Spills, U.S. Environmental Protection Agency, Louisville, KY, May 13-15, 1980.

Wagner, O.R. and Leach, R.O., Effect of interfacial tension on displacement efficiency, *Soc. Pet. Eng.*, 6:335-44, 1966.

Williams, D.E. and Wilder, D.G., Gasoline pollution of a groundwater reservoir—a case history, *Groundwater*, 9:50-54, 1971.

CHAPTER 25

Adsorption of Ionic Surfactants and Their Role in Colloid Stability

D.W. Fuerstenau

Both the polar group and the hydrocarbon chain on ionic surfactants enter into their interaction with minerals and colloids in aqueous media. The nature of the polar group determines the type of interaction with the surface, which may be either by physical adsorption as counter ions in the double layer, chemical adsorption, or even surface precipitation. The structure of the hydrophobic part of the molecule governs surfactant association processes in solution, as well as on the surface. Monomers dominate at low surface coverage, but at higher adsorption densities, they aggregate to form hemimicelles through hydrocarbon chain association. Charge reversal due to hemimicelle formation can enter into the destabilization (flocculation) and restabilization (dispersion) of colloidal suspensions. Under certain conditions, interaction hydrocarbon chains on adsorbed colloids can cause suspension destabilization even though the particles are highly charged. Some examples of surfactant adsorption and colloidal dispersion are briefly treated.

CHAPTER 26

Reactions of Anionic Surfactants in Saturated Soils and Sediments: Precipitation and Micelle Formation

Chad T. Jafvert and Janice Heath

The effectiveness of or potential for contaminant removal from soils, sediments, or aquifer materials (collectively termed "media") by surfactants largely depends on the magnitude of the sorption reactions of the pollutants to the media, the magnitude of pollutant solubilization by surfactant micelles or monomers, and the magnitude of surfactant loss to the media by sorption or precipitation reactions. Considerable work has been published regarding pollutant-medium interactions. Similarly, the solubilization of various pollutants by several types of surfactants has been investigated quantitatively. Not receiving much attention are the interactions of surfactant monomers and micelles with media components at surfactant concentrations above the critical micelle concentrations (cmc), where enhanced solubilization is most significant.

In this study, therefore, we have examined the interactions of the anionic surfactant dodecylsulfate (DS) with sediment components at DS concentrations in the range of the cmc. At aqueous phase DS concentrations approaching the cmc as described herein, precipitation reactions involving divalent cations become important. Also, because pollutant solubilization by surfactant micelles can be orders of magnitude greater than that by surfactant monomers, it is necessary to distinguish between these two forms to quantitatively evaluate results of partition experiments. This distinction is nontrivial in natural systems because of the dependence on ionic composition in determining the cmc and, hence, the monomer concentration for anionic surfactants.

Mechanistic or semi-empirical equations have been developed to describe these phenomena in well characterized aqueous solutions (Stellner and Scamehorn 1989a, 1989b; Kallay et al. 1985; Kallay et al. 1986; Bozic et al. 1979; Fan et al. 1988). These equations account for surfactant pre-cipitation, micelle formation, counter-ion binding to micelles, and the necessary material balances. The constants related to each of these phe-nomena have been determined for DS in aqueous solutions containing sodium and/or specific divalent salts (Stellner and Scamehorn 1989a, 1989b; Kallay et al. 1985; Kallay et al. 1986). Two studies (Stellner and Scamehorn 1989a; Kallay et al. 1986) specifically address "hardness" (or divalent cation) tolerance of DS solutions with consideration given to the monovalent electrolyte (sodium) composition. In both these studies, how-ever, the hardness tolerance of the individual divalent cation was consid-ered separately.

To characterize these phenomena in sediments, soils or aquifers, the significant reactions that involve both magnesium and calcium species must be considered simultaneously. Such a system of equations easily follows from these previous studies (Stellner and Scamehorn 1989a, 1989b; Kallay et al. 1986). Therefore, we are able to quantitatively compare our experimental results of DS precipitation and micelle formation in sediment and soil suspensions with calculated results based on a series of equations and associated constants derived from precipitation boundaries of individu-al divalent cations in well-defined aqueous solutions. Experimental and calculated precipitation boundaries are given in which total aqueous phase DS is the independent variable, and total aqueous phase calcium is the dependent variable.

DS molecules may exist in aqueous sediment or soil slurries in several states or phases, including ionic monomers in solution ($[DS^-]_w$), micellar aggregates ($[DS^-]_{mic}$), precipitates such as $Ca(DS)_2$, monomers adsorbed to separable solids ($[DS^-]_s$), and other less well defined or easily characterized states, such as adsorbed monolayers or dimers or other aggregates smaller than micelles (Attwood and Florence 1983). To develop a reasonable modeling framework, the assumption is made that only the former three states are significant in sediment or soil suspensions at DS concentrations approaching the cmc. Additionally, we assume that the only ionic species affecting the cmc and binding to micelles as counter-ions are sodium, calcium, and magnesium. Also, although all three of these ions can form precipitates with DS (based on their respective solubility products and concentrations in most freshwater systems), we assume that the calcium ion concentration controls the formation of precipitates.

Given these assumptions, the equations given in Table 1 follow directly from previous studies on DS speciation in well-defined aqueous solutions (Stellner and Scamehorn 1989a, 1989b; Kallay 1985). The equations are only briefly described here because more thorough descriptions are given by Stellner and Scamehorn (1989a, 1989b), and Kallay et al. (1986). Equations (1) through (4) of Table 1 are the material balances on total aqueous phase concentrations of DS, Na, Ca, and Mg, where the subscripts

Table 1. Reactions involving dodecylsulfate and medium components

$$[DS^-]_{aq} = [DS^-]_w + [DS^-]_{mic} \tag{1}$$

$$[Na^+]_{aq} = [Na^+]_w + [Na^+]_{mic} \tag{2}$$

$$[Ca^{2+}]_{aq} = [Ca^{2+}]_w + [Ca^{2+}]_{mic} \tag{3}$$

$$[Mg^{2+}]_{aq} = [Mg^{2+}]_w + [Mg^{2+}]_{mic} \tag{4}$$

$$[DS^-]_{mic} = [DS^-]_{aq} - cmc \qquad ([DS^-]_{aq} > cmc) \tag{5a}$$

$$[DS^-]_{mic} = 0.0 \qquad ([DS^-]_{aq} \leq cmc) \tag{5b}$$

$$K_{sp,Ca} = \{Ca^{2+}\}_w \{DS^-\}_w{}^2 = [Ca^{2+}]_w [DS^-]_w{}^2 \gamma_{Ca} \gamma_{DS}{}^2 \tag{6}$$

$$I = 0.5 \sum z_i^2 c_i \approx 3.0 ([Ca^{2+}]_{aq} + [Mg^{2+}]_{aq}) + [Na^+]_{aq} \tag{7}$$

$$\log \gamma_i = 0.5 z_i \left(\frac{\sqrt{I}}{\sqrt{I} + 1} - 0.3 I \right) \tag{8}$$

$$K_{Ca} = \frac{2 [Ca^{2+}]_{mic} [Na^+]_w \gamma_{Na}}{[Na^+]_{mic} [Ca^{2+}]_w \gamma_{Ca}} \tag{9}$$

$$K_{Mg} = \frac{2 [Mg^{2+}]_{mic} [Na^+]_w \gamma_{Na}}{[Na^+]_{mic} [Mg^{2+}]_w \gamma_{Mg}} \tag{10}$$

$$0.65 [DS^-]_{mic} = [Na]_{mic} + 2 [Ca^{2+}]_{mic} + 2 [Mg^{2+}]_{mic} \tag{11}$$

$$ln \ cmc = K_3 + K_g \ ln \ P \tag{12}$$

$$K_2 S^2 = [Na^+]_w P + ([Ca^{2+}]_w + [Mg^{2+}]_w) P^2 \tag{13}$$

$$S = \frac{[Na^+]_w P S_{Na}}{CMC_{SDS} \ P_{SDS}} \quad \frac{([Ca^{2+}]_w + [Mg^{2+}]_w) P^2 S_{Ca}}{0.5 \ CMC_{CDS} \ (P_{CDS})^2} \tag{14}$$

$$K_g = \frac{[Na^+]_w P K_{gNa}}{CMC_{SDS} \ P_{SDS}} + \frac{([Ca^{2+}]_w + [Mg^{2+}]_w) P^2 K_{gCa}}{0.5 \ CMC_{CDS} \ (P_{CDS})^2} \tag{15}$$

refer to species in either the water-dissolved phase (w), the micellar or micellar-bound phase (mic), or the sum of these components as the total aqueous phase (aq). The concentration of dodecylsulfate in micelles, $[DS^-]_{mic}$, is calculated from either Eq. (5a) or (5b) of Table 1, depending on the validity of the associated constraint.

Equation 6 is the solubility product for calcium dodecylsulfate. Although this equation appears rather straightforward, the calculation of the activity coefficients in solutions containing micelles can be somewhat problematic (see Burchfield and Woolley 1984). Fortunately, model sensitivity does not rely strongly on an explicit calculation of ionic strength, and a reasonable estimate can be made based on the major cations in solution. In this study, ionic strength I is calculated from total aqueous concentrations of Na, Ca, and Mg, assuming all anions in solution are monovalent species according to Eq. (7). Using this value, the activity coefficients can be calculated with the Davies equation, Eq. (8) (Davies 1962).

We have expanded the approach taken by Kallay et al. (1986) and have modeled competitive binding on micelles as ion exchange equilibria, according to Eqs. (9) and (10) of Table 1, where K_{Ca} and K_{Mg} are ion-exchange constants for calcium and magnesium ions, respectively, both referenced to exchange of sodium ion. The charge balance on micellar DS, given by Eq. (11), accounts for less than unity charge neutralization on micelles by assuming the total fractional charge neutralized by sodium, calcium, and magnesium ions is 0.65 (Stellner and Scamehorn 1989a; Rathman and Scamehorn 1987).

A set of equations and the corresponding coefficients defining the cmc have been developed by Stellner and Scamehorn (1989a) based on the earlier theoretical considerations of Shinoda (1955, 1963) and Moroi et al. (1974, 1975). We restate these equations in Table 1, assuming the effect on the cmc by Ca and Mg ions is the same. The parameters S_{Na}, S_{Ca}, CMC_{SDS}, CMC_{CDS}, P_{SDS}, P_{CDS}, $K_{g,Na}$, $K_{g,Ca}$, K_2, and K_3 are constants and cmc, S, P, and K_g are parameters that can be evaluated simultaneously if $[Na^+]_w$, $[Ca^{2+}]_w$, and $[Mg^{2+}]_w$ are known. Values for the constants, taken directly from Stellner and Scamehorn (1989a, 1989b), are given in Table 2. Abbreviations for the constants are identical to those in the cited reference. Equation (12) is a derivation of an equation developed by Shinoda (1955) relating the cmc to electrical potential, where P is a function of electrical potential at the surface. Equation (13) is derived from the Poisson-Boltzman equation, where S represents micellar surface charge. Equation (14) is a linear combination of the individual species contributions to the surface charge density on the micelle. The contribution of each species is normalized using the constants P_{SDS}, P_{CDS}, CMC_{SDS}, and CMC_{CDS}, determined from experimental data at DS concentrations at which micelles begin to form (i.e., < 2 % of DS is micellar). This DS concentration is the classical definition of the cmc.

Table 2. List of constants and variables

Knowns	Value	Unknowns
$K_{sp,Ca}$	5.0×10^{-10} M^3	$[DS^-]_{mic}$
K_2	2.87×10^5 M m^4 C^{-2}	$[DS^-]_{mon}$
K_3	-10.54	cmc
$K_{g,Na}$	0.698	$[Ca^{2+}]_w$
$K_{g,Ca}$	0.895	$[Ca^{2+}]_{mic}$
S_{Na}	7.96×10^{-3} C m^{-2}	$[Mg^{2+}]_w$
S_{Ca}	6.82×10^{-3} C m^{-2}	$[Mg^{2+}]_{mic}$
CMC_{SDS}	6.65×10^{-3} M	$[Na^+]_w$
CMC_{SDS}	1.90×10^{-3} M	$[Na^+]_{mic}$
P_{SDS}	2.74×10^3 M	γ_1
P_{CDS}	1.19×10^2 M	γ_2
K_{Mg}	32.0	I
K_{Ca}	63.0	P
$[DS^-]_{aq}$		S
$[Mg^{2+}]_{aq}$		K_g
$[Na^+]_{aq}$		$[Ca^{2+}]_{aq}$

All values are from Stellner and Scamehorn (1989a, 1989b),
except for K_{Mg} and K_{Ca}, which are from Kallay et al. (1986).

Numerical values for all constants contained in the equations are given
in Table 2. In total, there are 16 equations (γ_i must be evaluated at z = 1
and 2, and $\gamma_{Ca} = \gamma_{Mg}$) and 19 additional parameters. To evaluate all parameters, values for an additional three must be known. From batch experi-
ments, values of $[DS^-]_{aq}$, $[Ca^{2+}]_{aq}$, $[Mg^{2+}]_{aq}$, and $[Na^+]_{aq}$ are easily
determined. This allows us to use three of these as input parameters to
calculate all remaining unknown values, including the fourth (measured)
parameter. We have used $[DS^-]_{aq}$, $[Mg^{2+}]_{aq}$, and $[Na^+]_{aq}$ as input parameters

to calculate precipitation boundaries on $[Ca^{2+}]_{aq}$, which can be compared with the experimental $[Ca^{2+}]_{aq}$ data. The calcium concentration was selected as the dependent variable because it is more sensitive to $[DS^-]_{aq}$ than was magnesium or sodium concentration. The equations are solved by iteration.

Our experiments involved the addition of varying total amounts of DS to a series of tubes containing constant amounts of a medium. Recoveries after 24 h incubations of aqueous phase DS are shown in Figure 1 for six different sediments or soils (media). The media used in these experiments were those characterized by Hassett et al. (1980). Aqueous DS was determined by the method developed by Motomizu et al. (1982), with a modification in reagent volumes made to accommodate 15 mL test tubes. In these experiments, the medium mass to aqueous volume ratio (m/v) was held at 0.2 g/mL. For each medium, basically two straight lines describe the experimental data; the first line extending from the intercept at $[DS^-]_{aq}$ = 0.0 to ≈5 mM $[DS^-]_{aq}$, and the second line extending from this point with a slope of ≈1. For media 19 and 22, this change in slope is not as obvious, because recovery of DS in the aqueous phase is generally > 75%. For media 11 and 20, however, recovery of DS is sometimes < 25 %, and a clear break in the recovery line exists.

The cmc of the sodium salt of DS is generally accepted to be ≈8.2 mM (Mukerjee and Mysels 1971). Divalent cations in solution will decrease the cmc below this value. As will be shown, the breakpoint on each data set occurs at the value of $[DS^-]_{aq}$ where micelles begin to form. Because the slope of DS recovery beyond this breakpoint is ≈1 for all the media, the differences in recovery among the media is dependent on the cmc value and the slope of the recovery line below the cmc, where all DS is monomeric.

If precipitation is occurring in these samples, the aqueous calcium concentration should decrease as $[DS^-]_{aq}$ increases (i.e., at values of $[DS^-]_{aq}$ where no micelles are present with a negative slope equal to the activity-corrected solubility product). As $[DS^-]_{aq}$ increases beyond the initial point of micelle formation, counter-ion binding of calcium to micelles should increase the aqueous calcium concentration. These phenomena are evident in Figure 2, where aqueous calcium, magnesium, and sodium concentrations are plotted against aqueous DS for medium 16. We have also shown on this figure the calculated calcium precipitation boundaries, obtained by solving Eqs. (1) through (15) with the constants described in Table 2 and the experimental values of $[DS^-]_{aq}$ and $[Mg^{2+}]_{aq}$, and $[Na^+]_{aq}$ for this experiment.

The calculated calcium concentrations agree favorably with the experimental values, except at very low $[DS^-]_{aq}$ for all media and along the solubility product line for media 11 and 16. The discrepancies at low concentrations, as shown in Figure 2 for medium 16, result from aqueous

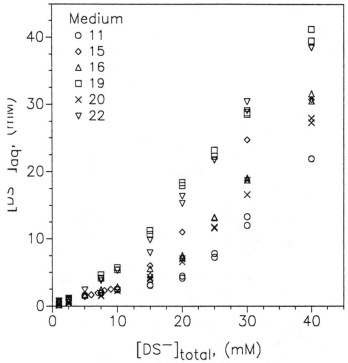

Figure 1. Aqueous phase DS recovery from 0.2 g/mL slurries of the media after 24 h, as a function of the amount of DS added to each tube (referenced to the aqueous phase volume).

DS and calcium concentrations less than those required to cause precipitation. On the other hand, several factors may account for the discrepancies between the calculated and experimental solubility product lines for medium 16. Possibly, either a fraction of the aqueous calcium is chelated to natural dissolved materials, or solution heterogeneity results in thermodynamically less stable precipitates than those formed in the homogeneous solutions from which the $K_{sp,Ca}$ of 5×10^{-10} M^3 was measured. Other investigators have determined $K_{sp,Ca}$ to be as low as 2.14×10^{-10} M^3 in pure solution (Kallay et al. 1986). In any case, even for medium 16, the experimental conditional solubility products of calcium dodecylsulfate are within a factor of 2 of the measured value of 5×10^{-10} M^3.

Calculating the fraction of DS existing as either monomers or micelles follows from the solution of Eq. (1) through (15). For medium 16 ($m/v = 0.2$ g/mL), the addition of over 10 mM of DS is required to initiate micelle formation at an aqueous concentration of ≈ 4.0 mM. At higher $[DS^-]_{aq}$, the

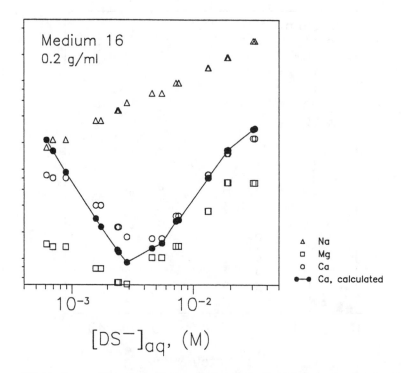

Figure 2.   Aqueous phase concentrations of inorganic species as a
function of aqueous DS concentration in 0.2 g/ml medium
16 slurries. Solid symbols connected by solid lines represent calculated calcium concentrations.

calculations suggest that the concentration of DS monomers decreases; at
31 mM [DS$^-$]$_{aq}$, the monomer concentration is 2.43 mM.

The aqueous calcium concentrations associated with the soils and
sediments examined in this study (at m/v = 0.2 g/mL) are in the range of
values found for most natural surface and ground waters. Therefore, the
results presented indicate that, at DS surfactant doses approaching the cmc,
the predominant mechanism of DS loss to most natural soils, sediments
and aquifer material will be precipitation.

Stellner and Scamehorn (1989a, 1989b) and Kallay et al. (1985, 1986)
have modeled the hardness tolerance of DS in well-defined aqueous solutions. We have applied their equations and associated constants for the
precipitation of Ca(DS)$_2$, for the binding of counter-ions on the micelle
surface, and for the calculation of the cmc. This allows for the determination of the fraction of surfactant in micellar and monomer form and an

estimation of the surfactant dose necessary to obtain a given micellar concentration. This distinction becomes critical in quantitatively evaluating results of enhanced solubility experiments.

Processes unaccounted for in this analysis are exchange reactions involving aqueous- and medium-associated cations. Inclusion of such reactions would allow a prior determination of total aqueous surfactant concentrations, but other experimental information also would be required. It should be noted, however, that the equilibrium between micellar- and medium-associated species must be mediated through the free solution phase concentration. For example, micellar- and medium-associated calcium are both in equilibrium with $[Ca^{2+}]_w$. Because the calculated value for $[Ca^{2+}]_w$ does not change appreciably for any DS concentration where micelles are present, the relative amount of calcium associated with the micelle and media phases should be primarily dependent on the relative amount of each phase. This appears to be the case resulting in precipitation diagrams that resemble those predicted from solution phase theories.

ACKNOWLEDGMENT

The authors wish to acknowledge the helpful comments of Patricia Van Hoof.

REFERENCES

Attwood, D. and Florence, A.T., *Surfactant Systems,* Chapman and Hall, New York, 1983.

Bozić, J., Krznarić, I., and Kallay, N., *Colloid Polymer Sci* 257:201-05, 1979.

Burchfield, T.E. and Woolley, E.M., *J Phys Chem* 88:2149-55, 1984.

Davies, C.W., *Ion Association,* Butterworths, London, 1962.

Fan, X., Stenius, P., Kallay, N., and Matijević, E., *J Colloid Interface Sci* 21:571-78, 1988.

Hassett, J.J., Means, J.C., Barnwart, W.L., and Woods, S.G., Sorption properties of sediments and energy-related pollutants. EPA 600/3-80-041 U.S. Environmental Protection Agency, Athens, GA, 1980.

Kallay, N., Pastuović, M., and Matijević, I., *J Colloid Interface Sci* 106:452-58, 1985.

Kallay, N., Fan, Xi-Jing, and Matijević, E., *Acta Chemi Scandinavica* A40:257-60, 1986.

Moroi, Y., Motomura, K., and Matuura, R.J., *J Colloid Interface Sci* 46:111, 1974.

Moroi, Y., Nishikido, N., and Matuura, R.J., *J Colloid Interface Sci* 50:344, 1975.

Motomizu, S., Fujiwara, S., Fujuwara, A., and Toel, K., *Anal Chem* 54:393-94, 1982.

Mukerjee, P., and Mysels, K.J., Critical micelle concentrations of aqueous surfactant systems, NSRDS-NBS 36. U.S. Department of Commerce, National Bureau of Standards, Washington, DC, 1971.

Rathman, J.F. and Scamehorn, J.F., *Langmuir* 3:372-77, 1987.

Shinoda, K., *Bull Chem Soc Japan*, 28:340, 1955.

Shinoda, K., *Colloidal Surfactants*, Shinoda, K., Tamamushi, B., Nakagawa, T., and Isemura, T., (eds.), Academic Press, New York, 1963, pp 39-42.

Stellner, K.K. and Scamehorn, J.F., *Langmuir* 5:70-77, 1989a.

Stellner, K.K. and Scamehorn, J.F., *Langmuir* 5:77-84, 1989b.

CHAPTER 27

Soil and Clay Modification for Environmental Restoration and Waste Management

Stephen A. Boyd

The sorptive capabilities of clays for nonionic organic contaminants (NOCs) can be greatly improved by replacing naturally occurring inorganic exchange cations (e.g., CA^{2+}, Na^+) with organic cations of the general form $[(CH_3)_3NR]^+$, where R may be an alkyl or aromatic hydrocarbon. Organic cations such as hexadecyltrimethylammonium (HDTMA, R - $C_{16}H_{33}$) will nearly stoichiometrically displace inorganic exchange ions of natural clays by simple ion exchange reactions. This results in the formation of an organic phase, derived from the interaction of the alkyl moieties (R groups), that is fixed on the clay surfaces and interlayers. These organic phases act as powerful partition media for NOCs and effectively remove such compounds from water. Organic matter contents of 10% to 30% (w/w) or more can be obtained by exchanging 2:1 clay minerals with HDTMA. The K_{om} values of these organo-clays are similar to each other and to the corresponding K_{ow} values, showing that HDTMA forms a highly effective partition phase when exchanged on smectite, illite and vermiculite. The NOC sorption coefficients of HDTMA clays can be increased (up to five times) by using clays with high surface charge densities.

Soils and subsoils exchanged with HDTMA also show greatly enhanced sorption capabilities for NOCs. Sorption coefficients for NOCs in the modified soils increased by over two orders of magnitude. The K_{om} values of HDTMA-exchanged low- and high-clay content soils were similar to each other suggesting the general applicability of this soil modification technique. These K_{om} values were also similar to the corresponding K_{ow} values demonstrating that the added HDTMA organic matter was about 10 to 30 times more effective than natural soil organic matter as a partition phase for NOCs. These results demonstrate the capability of forming

177

organ-clays derived from soil clays, and provides the basis for an effective in situ approach for increasing the assimilative capacity of soils for NOCs.

When smaller organic cations, such as trimethylphenylammonium (TMPA), are used as exchange ions on smectite clays, the resulting organo-clays behave as surface adsorbents. Trimethylphenylammonium-smectite was shown to be a highly effective adsorbent for the water soluble aromatic constituents of petroleum, i.e., benzene, toluene, xylene, ethylbenzene, butylbenzene and naphthalene. The use of high-charge smectites decreased the effectiveness of the naphthalene. The use of high-charge smectites decreased the effectiveness of the resultant organo-clays. Trimethylphenylammonium-smectite was the most effective adsorbent of aromatic hydrocarbons among a related series of organo-clays that included tetramethyl-, benzyltrimethyl, benzyltriethyl-, and benzyltributylammonium.

These results show that organo-clays are promising fundamental materials for use in a variety of environmental applications including: chemical waste stabilization, as components of clay barriers for hazardous waste storage reservoirs, as liners for petroleum storage tanks, and in the development of in situ restoration technologies for contaminated soils and subsoils.

SUMMARY AND RESULTS OF

SECTION IV

MODIFIED COLLOIDS, SURFACTANTS, AND EMULSIONS

CHAPTER 28

Discussion Leaders

John C. Westall and Richard G. Luthy

The group identified five strategies and these are outlined in Table 28.1 (page 186). The two most feasible strategies are discussed in more detail below.

APPROACH 1: MOBILIZE CONTAMINANTS VIA SURFACTANTS

Description

Two fundamental approaches are foreseen: (i) extract organic and inorganic pollutants into mobile micelles, microemulsions, and emulsions; (ii) promote immiscible displacement of organic liquids via alteration of surface tension at the aquifer pore/entrapped organic liquid/water interface. A related area is the delivery of chemical treatment (e.g., oxidants or reductants) in micelles, which protect the reactant from reaction with water or the environmental matrix.

Processes

Three levels of processes are recognized: (i) the effect of surfactants on the phase behavior of water, organic liquids, (and the surfactants themselves, if they are introduced as a separate phase); (ii) the effect of surfactants within a phase on interfacial tension and wetting at interfaces between environmental sorbents, water, and organic liquids; and (iii) the effect of surfactants on the distribution of organic and inorganic pollutants among the various phases.

Advantages

This approach is flexible with a wide range of options. Field experience exists from enhanced oil recovery and solution mining for uranium.

Limitations

While no fundamental limitations of the method were recognized, many unresolved questions, which depend on site-specific parameters, arose. The questions are listed under "gaps in knowledge."

Gaps in Knowledge

General gaps in knowledge include (i) uncertainties about the processes and (ii) the relation between lab scale and field scale. Other questions might arise as specific applications are considered; it is difficult to speculate on the impact of these issues without considering specific sites. Some of these issues might turn out to be limitations of the approach at some sites, including:

- the extent of recovery desired: first 50% or last 5%;
- secondary dispersion of particles by surfactants;
- charge inversion if too much surfactant is added;
- formation of mixed micelles;
- kinetics: transport slow, reaction fast;
- clogging of pores by surfactant;
- degradability of surfactant;
- fate of residual surfactant;
- extent of adsorption of surfactant on aquifer material;
- unavoidable formation of emulsions;
- treatability of the extracted pollutant.

Overall Rating

This approach was judged to be the most promising and the first priority for additional study. There exists a high chance of success. A pilot study could be designed based on what we know today.

APPROACH 2: ALTER PERMEABILITY OF MEDIUM SELECTIVELY TO PROMOTE DELIVERY OF "TREATMENT" (i.e., extractant or in situ reactant) TO PORES.

Description

In situ treatment processes can be ineffective due to the heterogeneity in hydraulic conductivity of the porous medium: the treatment tends to follow paths of high conductivity and miss the low-conductivity paths. To correct this problem, surfactants that are capable of forming gels or foams in situ would be pumped through the aquifer. The gels or foams would form in the high-conductivity channels, reducing their conductivity and forcing further treatment into the smaller pores, resulting in a more uniform treatment of the entire aquifer. The "treatment" that would then be delivered could be an extractant, an in situ chemical treatment, or an in situ biological treatment (organisms or nutrients).

Processes

The processes that are unique to this approach are (i) the kinetics of in situ gel and foam formation and (ii) the influence of gels and foams on viscosity. Several classes of surfactant materials would have to be investigated for suitability. Another process common to all in situ treatment procedures is transport of the treatment itself through the porous medium.

Advantages

This approach is flexible with a wide range of options. Field experience exists from enhanced oil recovery.

Limitations

While no fundamental limitations of the method were recognized, many unresolved questions, which depend on site-specific parameters, arose (see "gaps in knowledge").

Gaps in Knowledge

General gaps in knowledge include (i) uncertainties about the processes and (ii) the relation between laboratory scale and field scale. Other questions might arise as specific applications are considered; it is difficult to speculate on the impact of these issues without considering specific sites. Some of these issues might turn out to be limitations of the approach at some sites:

- choice of surfactants;
- the kinetics of transport vs the kinetics of gel and foam formation;
- the influence of gel or foam on hydraulic conductivity;
- adsorption of surfactant on aquifer material;
- unavoidable formation of emulsions;
- degradability of surfactant;
- fate of residual surfactant;
- treatability of the contaminant;
- the extent of treatment desired: first 50% or last 5%.

Overall Rating

This approach was judged to be very promising and certainly worthy of further consideration. There exists a high chance of success. A pilot study could be designed based current understanding.

DISCUSSIONS AND CONCLUSIONS

The group agreed that two approaches are very promising and warrant further study: (i) mobilization of pollutants with surfactants, and (ii) use of surfactants to modify aquifer permeability. Three other approaches were considered and handled as follows.

The use of surfactants to alter the stability of colloidal dispersions was considered to be very promising. Thus the mobility of natural or man-made colloids could be altered. However, it was felt that this topic belonged organizationally in the report on Session I.

The use of surfactants to modify clay surfaces to promote the sorption of hydrophobic compounds was considered. This approach was judged to be very promising, but some members of the group felt that it was outside the bounds of a program on colloid chemistry. Thus the group endorsed further work on the approach, but outside of the colloids program.

Another approach that was suggested was the prevention of in situ formation of colloidal particles through the use of surfactants to influence nucleation and crystal growth. While laboratory studies document the viability of this approach in clean systems, it was generally agreed that the number of field systems in which this approach might be applied (e.g., the formation of $Fe_3(PO_4)_2$ at Cape Cod) are small, and the transfer from the laboratory to the field might be particularly difficult. However, this idea might be kept in mind for application at some unique sites.

In conclusion, the group agreed that two approaches are very promising and warrant further study: (i) mobilization of pollutants with surfactants, and (ii) use of surfactants to modify aquifer permeability. These approaches are mechanistically simple, chances of success are high, and pilot studies could be designed with what we know today. Key research problems that remain include: (i) the relation between lab scale and field scale; and (ii) uncertainties about the processes:

- Phase behavior of surfactants/water/organic liquids
- Surface tension, wetting
- Distribution of contaminants between water, surfactant phase, and sorbents
- Selection of surfactants for formation of gels and foams
- Kinetics and equilibrium of gels and foams
- Influence of gels and foams on hydraulic conductivity of medium

TABLE 28.1 Evaluation of strategies considered in group discussions.

PROPOSED MANIPULATION APPROACH	KEY HYDRO-GEO-BIO-CHEMICAL PROCESSES IN APPROACH	ADVANTAGES/ REASONABLENESS OF APPROACH	LIMITATIONS/ IMPRACTICALITIES OF APPROACH	MAJOR GAPS IN KNOWLEDGE	OVERALL RATING
1. Mobilize pollutants via surfactants	Phase behavior of surfactants/ water/organic liquids Surface tension, wetting Distribution of organic pollutants Distribution of inorganic pollutants	Flexible, wide range of options	Biological degradation of surfactants Undesirable effects of residual surfactants	Lab scale vs. field scale Scale (placement of wells) Extent of adsorption of surfactant on aquifer material	High chance of success Pilot study could be designed with what we know today
2. Alter permeability of medium selectively to promote delivery of "treatment" (i.e., extractant or in situ reactant) to pores	Transport of raw material to site Formation of gels and foams Influence of gels and foams on viscosity	Flexible, wide range of options Experience from enhanced oil recovery	Biological degradation of surfactants Undesirable effects of residual surfactants	Processes Lab scale vs. field scale Long-term fate of material The best foams and gels	High chance of success Pilot study could be designed with what we know today
3. Alter the mobility of other colloids (natural or introduced intentionally)	Stability of colloids Filtration of colloids Adsorption of surfactants to colloid particles / porous matrix	Flexible, wide range of options	Biological degradation of surfactants Undesirable effects of residual surfactants	Processes	

PROPOSED MANIPULATION APPROACH	KEY HYDRO-GEO-BIO-CHEMICAL PROCESSES IN APPROACH	ADVANTAGES/ REASONABLENESS OF APPROACH	LIMITATIONS/ IMPRACTICALITIES OF APPROACH	MAJOR GAPS IN KNOWLEDGE	OVERALL RATING
4. Promote sorption of nonpolar organic compounds to mineral surfaces through the use of surfactants	Competition among ionic sorbates Hydrophobic interaction involving surfactant-modified surfaces	Flexible, wide range of options	Biological degradation of surfactants Undesirable effects of residual surfactants	Processes	
5. Prevent in situ formation of colloidal particles	Nucleation and crystal growth	Possibility to "sterilize" system with respect to generation of colloids	Possibility of many mechanisms leading to formation of colloids, with each mechanism responding in a different way to surfactants Biological degradation of surfactants Undesirable effects of residual surfactants	Processes	Low likelihood of success in field situations

SECTION V

ABSTRACTS OF
POSTER PRESENTATIONS

Polynuclear Aromatic Hydrocarbon (PAH) Binding to Natural Organic Matter (NOM): A Comparison of NOM Fractions

Gary L. Amy, Martha H. Conklin, Houmao Liu, and Christopher Cawein

It is well known that polynuclear aromatic hydrocarbons (PAHs) bind to humic substances. In this study, the binding of a model PAH, phenanthrene, to various natural organic matter (NOM) fractions has been investigated. Batch-mode experiments have been conducted to study the aqueous-phase system and the heterogeneous system involving alumina, a model mineral phase. Higher-molecular-weight (MW) and humic fractions of NOM preferentially bind PAHs. In our work, PAHs have shown a weaker binding affinity for mineral-bound (i.e., alumina-bound) as opposed to dissolved NOM. Fluorescence quenching has shown merit as an analytical technique to differentiate between free and NOM-bound PAH.

The movement of hydrophobic organic compounds in groundwater systems is controlled by their partitioning between the aqueous (mobile) and solid (immobile) phases. If there is another nonsorbing compound in the aqueous phase that interacts with the contaminant, the hydrophobic organic compound may preferentially remain in solution and its transport may be facilitated. The behavior of aquatic NOM falls within this context. The movement of the solute will be attenuated, however, if the interacting compound is sorbed on the aquifer media; the behavior of soil organic matter falls within this realm. One objective of this research was to identify important characteristics of aquatic NOM that affect partitioning and transport of PAHs. Another objective of the research was to delineate the roles of dissolved vs mineral-bound NOM in subsurface partitioning and transport.

In this work, the partitioning between one PAH, phenanthrene, and each of two sources of groundwater NOM has been studied in the presence and absence of a model mineral surface. The behavior of a soil-derived fulvic acid sorbed to the mineral surface has also been investigated.

This work has focused on phenanthrene, which has a log K_{ow} of 4.4 and a water solubility of 75 µg L^{-1}. We have used fluorescence for detecting free phenanthrene in solution.

Two sources of aquatic NOM have been evaluated: (1) groundwater derived from a basin underlying Orange County (OC), California, and (2) groundwater obtained from the Biscayne Aquifer (BA), Florida. We also have evaluated a soil fulvic acid as a model of mineral-bound NOM. Ultrafiltration (UF) was used to produce discrete MW fractions, resulting in permeates and retentates that were reconstituted back to original pH (8) and ionic strength (~0.01) conditions. XAD-8 resin adsorption was used to produce humic and nonhumic fractions of NOM; the humic fraction was reconstituted back to the original pH and ionic strength. The soil fulvic acid was prepared by dissolution 0.01 M solution of $NaHCO_3$ and adjustment to a pH of 8.0. The unfractionated, "bulk" NOM associated with the OC source exhibited a substantially higher percentage of humic material than the BA source and a slightly larger average MW.

Alumina, α-Al_2O_3 has been used as a model mineral phase. The average size was reported to be 1.85 µm, and the surface area was 6.5 m^2 g^{-1}. The presence of HCO_3^- has been found to influence its electrophoretic mobility (and pH_{zpc}): the pH_{zpc} ranged from 7.2 (with HCO_3^-) to 8.5 µm sec^{-1} per volt cm^{-1} (without HCO_3^-).

Several sets of experiments have been performed to determine the best method for determining free PAH concentrations and NOM-bound PAH concentrations in the aqueous phase. Batch experiments were performed to evaluate three techniques: fluorescence quenching, reverse-phase (Sep-Pak) separations, and dialysis bags. The NOM-bound PAH, PAH-NOM, is defined as the initial PAH_0 minus the PAH at equilibrium. Expressing the concentration of NOM as dissolved organic carbon (DOC), the binding constant, K_{doc}, is defined as

$$PAH + NOM \leftrightarrow PAH\text{-}NOM$$

$$K_{doc} = \frac{PAH - NOM}{(PAH)\,(NOM)}$$

The lowest recovery of PAH was observed for the dialysis technique, which indicates that these results are the most questionable. Better recovery was obtained with the Sep-Pak method; however, some losses were observed. Because FQ and Sep-Pak methods agreed within a factor of two, we selected the FQ method for subsequent experiments.

We have conducted batch experiments to study the three binary systems (PAH-NOM, PAH–α–Al$_2$O$_3$, and NOM–α–Al$_2$O$_3$) and the ternary system (PAH-NOM–α–Al$_2$O$_3$). The binding experiments reported herein have been based on a temperature of 23° C, a pH of ~8, and an ionic strength of ~0.01.

Binding constants have been measured for different MW and humic/nonhumic fractions of the two aqueous NOM sources. The following observations can be made: (1) larger binding constants were found for the OC source than the BA source, (2) the humic fraction for the OC and BA sources exhibited a higher binding constant than the bulk-NOM for these sources, and (3) the higher-MW fraction of each source exhibited a higher binding constant than did the lower-MW fraction. Overall, the humic/high-MW fraction of each NOM source exhibits the greatest binding properties.

Batch experiments have been performed to measure equilibrium isotherms for the sorption of NOM onto α–Al$_2$O$_3$ under bulk NOM source conditions (pH ~8). The effects of NOM-coating on surface chemistry are highlighted by the change in electrophoretic mobility of the particles. As expected, the NOM coating increased the negative surface charge of the alumina particles. This binary system is intended to serve as a model of mineral-bound NOM.

A single point isotherm experiment was run to determine the potential for PAH binding directly to the mineral surface in the complete absence of NOM. We found a solid-phase concentration of 8.8 µg g^{-1} α–Al$_2$O$_3$ in equilibrium with an aqueous-phase concentration of 66 µg L^{-1}, corresponding to an equivalent linear partition coefficient of 133 cm^{3}g^{-1}.

Relatively little is known about the chemical differences between aquatic NOM and soil NOM (i.e., soil organic matter) in groundwater systems. Intriguing questions exist about their genesis and interrelationships. Aquatic NOM can enter the subsurface via recharge or by in situ leaching of organic rich strata. As it follows advective flow, aquatic NOM comes into intimate contact with mineral matter, and sorption onto the mineral surface may occur if localized equilibrium does not exist. Thus, soil organic matter may originate as either aquatic NOM or as organic rich strata created through sedimentary processes.

Experiments that use the NOM sources have been conducted to study the partitioning of a PAH into α–Al$_2$O$_3$-bound NOM. In these experiments, a known mass of α–Al$_2$O$_3$ was contacted with a known volume of a NOM solution for an established equilibration time at each of several pH conditions. The mixture was centrifuged and rinsed with a NaHCO$_3$/NaCl solution (I = 0.01) at each of the pH conditions; although some desorption of

NOM was observed at the higher pH levels, virtually no desorption was seen at the lower pH levels. The NOM-coated solids were then contacted with a phenanthrene/NaHCO$_3$/NaCl solution (I = 0.01) at each of the corresponding pH levels, and the mixture was centrifuged after equilibration. Based on measurements of initial and final phenanthrene, estimates of PAH partitioning into the mineral-bound NOM were made. Thus, the overall protocol involved (1) first contacting bulk NOM with alumina to produce residual NOM and mineral-bound NOM and (2) then separately contacting residual NOM and mineral-bound NOM with PAH to evaluate binding. The overall process is described below:

$$NOM + 0\text{-}Al_2O_3 \leftrightarrow NOM + NOM\text{-}\alpha\text{-}Al_2O_3$$

$$NOM\text{-}\alpha\text{-}Al_2O_3 + PAH \leftrightarrow PAH\text{-}NOM\text{-}\alpha\text{-}Al_2O_3$$

$$NOM + PAH \leftrightarrow PAH\text{-}NOM$$

We determined binding constants, K_{doc}, for various NOM sources/ fractions at each of the pH conditions studied, and observed the following: (1) no binding occurred between the PAH and mineral-bound NOM for either aquatic NOM source, (2) less binding occurred between the PAH and residual NOM than that with the original bulk NOM, and (3) less binding occurred between the PAH and mineral-bound NOM than that between the PAH and bulk NOM for the soil fulvic acid. It is noteworthy that binding of a PAH with mineral-bound NOM depends on (1) the source of NOM, (2) surface coverage of NOM on the mineral surface, and (3) the configuration of NOM on the mineral surface (i.e., steric orientation, monolayer vs multilayer coverage, etc.). The observation of no binding between the PAH and mineral-bound NOM from both aquatic sources may result from (1) very low NOM coverage on the mineral surface, with the weak PAH binding properties of the bare mineral surface controlling, and/or (2) both aquatic NOM sources actually representing "residual" NOM that was already in equilibrium with aquifer media at the actual field sampling site(s), thus suggesting that aquatic NOM may be a poor model for soil organic matter.

Although the bulk NOM solution produced from the soil fulvic acid exhibited the largest binding constant, its binding properties were significantly reduced after its sorption onto the alumina surface. The observation of less binding of PAH with mineral-bound soil fulvic acid may be the results of changes in the configuration of the NOM when it adsorbs to the mineral surface. The observation that residual NOM exhibits less PAH binding than the corresponding bulk NOM is consistent with the premise that higher-MW NOM is preferentially adsorbed to the alumina surface.

Estimates of PAH solid-phase loadings indicate that, despite the lower binding properties of mineral-bound vs solution-phase NOM, the NOM-amended surface exhibits greater binding than the "bare" mineral (alumina) surface.

The relative abundance of dissolved vs mineral-bound NOM in a groundwater aquifer as well as differences in binding properties will influence the subsurface transport of PAHs.

ACKNOWLEDGMENT

This research has been supported by the Subsurface Science Program of the Ecological Research Division, Office of Health and Environmental Research, U.S. Department of Energy, Washington, D.C. (Program Manager: Dr. F.J. Wobber).

CHAPTER 30

Factors Influencing Nonaqueous Phase Liquid Removal from Groundwater by the Alcohol Flooding Technique

G.R. Boyd and K.J. Farley

Leaks of gasoline and industrial solvents from underground storage tanks pose serious threats to subsurface water supplies. The hydrocarbon mixtures that constitute these spills often exhibit low aqueous solubilities and persist in groundwater as separate, immiscible phases. These liquids are known as non-aqueous phase liquids (NAPLs). NAPL contaminants denser than the aqueous phase are referred to as DNAPLs.

Pump-and-treat remediation practices are currently employed to displace NAPL contaminants from groundwater. This remediation strategy removes only a fraction of the total contamination because of the trapping of NAPL globules by capillary forces. Trapped globules can occupy 14 to 30% (Wilson et al. 1990) of the total pore volume. Continued waterflooding only reduces the NAPL saturation by the slow process of dissolution. Complete contaminant removal often requires years or decades.

In situ volatilization has been successful for remediation of volatile organic contaminants in the unsaturated zone. However, this process may not be feasible for DNAPLs that penetrate deep into the saturated zone. Bioremediation programs can remove water-soluble and highly biodegradable contaminants, but the process tends to be less efficient for recalcitrant DNAPL contaminants. New technologies are therefore needed to expedite remediation processes without imposing undue harm to the environment.

One of the more promising remediation technologies involves the mobilization of residual NAPL globules by the injection of isopropyl alcohol (2-propanol). As isopropyl alcohol (IPA) is injected into the contaminated zone, interfacial tension (IFT) between the NAPL globule and the aqueous phase is reduced. This results in a reduction in the contact angle between the phases and elongation of the NAPL globules. The globules are then easily displaced through pore channel constrictions. Under certain conditions, the IFT between the two fluid phases may be eliminated and the NAPL is carried out with the solvent as a single phase.

Alcohol flooding has been shown to be technically feasible in the petroleum industry (Gatlin and Slobod 1960), and preliminary column experiments indicate that IPA may be suitable for NAPL removal from groundwater (Boyd and Farley 1990). The advantage of this technique is the ability of IPA to remove simultaneously NAPL residuals and adsorbed contaminants, thus reducing total remediation time. IPA is readily biodegradable and capable of reducing the IFT of both light and dense NAPLs. The process may be applied to both confined and phreatic aquifers.

The objective of our studies is to assess the merits of using IPA for DNAPL removal from groundwater. Evaluations of DNAPL behavior are performed using packed laboratory columns. As discussed below, column similitude is an important consideration to ensure that laboratory models properly represent field conditions.

The simultaneous flow of multiple fluid phases through porous media is governed by capillary, viscous, and gravitational forces. The influence of each of these forces was scaled in the laboratory to establish a model representative of field conditions. Two scaling studies were performed: (1) to study the effects of capillary forces, also known as end effects and (2) to study the effects of viscous and gravitational forces, also known as fingering effects.

Kyte and Rapoport (1958) investigated the operational parameters affecting the magnitude of residual saturation when NAPLs were displaced from a water-wet formation in the absence of gravitational forces. The investigators showed that scaling dimensions depend on the column length (L), the displacing fluid injection velocity (V), and viscosity (μ_d). The operating parameters that control capillary forces in the laboratory model are characterized by a scaling coefficient, $LV\mu_d$.

The investigators reported that as the scaling coefficient was increased, the amount of NAPL recovered from the column also increased. Above a critical value for $LV\mu_d$, known as the critical scaling coefficient, a maximum amount of NAPL was recovered, and the model most closely approximated field conditions. This operating configuration was termed "capillary stabilized flow."

In the absence of capillary forces, the macroscopic displacement efficiency of one fluid phase by another will be influenced by viscous and gravitational forces (Collins 1961). A perturbation analysis was performed to predict the influence of these forces on the displacement efficiencies of water, IPA, and a representative DNAPL, trichloroethylene (TCE). Results show that the injection velocity required for stable displacement depends on the density difference between the two fluids, the viscosity ratio, and the flooding orientation. Stability criteria for 0.5-mm diam glass beads in one-dimensional flow for the displacement processes in these studies are summarized in the following:

Displacement processes	Up flow	Horizontal flood	Down flow
H_2O Displacing TCE	V > 266 m/d	stable[a]	stable
IPA Displacing H_2O	V > 78 m/d	stable[a]	stable
H_2O Displacing IPA	V < 27 m/d	unstable	unstable

[a] Gravity segregation may occur under field conditions.

For our laboratory studies, experiments were conducted using a glass column (2.54 cm ID) packed to a total length of 80 cm with 0.5-mm-diam glass beads. Injection velocities were maintained at 5.5 m/day. The column dimensions and operating conditions represented a scaling coefficient ($LV\mu_d$ = 30 cm^2·cp·min^{-1}) greater than the critical value required for stabilized capillary flow.

Experimental results are shown in Figures 1 and 2 for TCE recovery by waterflooding and then alcohol flooding in the down-flow direction (see table). Figure 1 shows TCE saturation in the column as a function of cumulative fluid injection. The column was initially saturated to 69% and the free product was displaced by waterflooding, leaving a residual TCE saturation of 12%. A 1 pore volume (PV) slug of IPA was introduced, and within 1 PV of fluid injection, the TCE saturation was reduced to zero.

Figure 2 shows TCE and IPA concentrations measured at the column effluent commencing at the time of IPA injection. After ~0.4 PV of injection, TCE concentrations increased suddenly from aqueous solubility to maximum concentration. This occurred prior to the appearance of IPA in the effluent indicating the mobilization of residual TCE as a separate phase. As the IPA concentration increased, the effluent samples evolved from a milky emulsion into a single-phase solution, whereby the TCE was completely dissolved in the alcohol phase. Within 2 PVs of IPA injection, TCE was reduced to nondetectable concentrations. Figure 2 also shows that approximately 6 PVs of subsequent water flushes were required to reduce the IPA concentration to a nondetectable level. This inefficient IPA displacement was attributed to the unfavorable fluid injection sequence (i.e., H_2O displacing IPA in the down-flow mode is expected to be unstable).

Similar results were obtained in another experiment in which waterflooding and alcohol flooding were conducted in the horizontal direction.

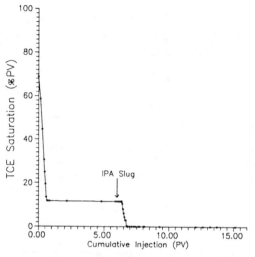

Figure 1.   TCE saturation versus cumulative fluid injection for a column operated in the down flow direction.

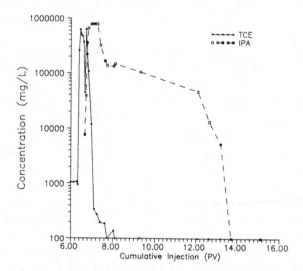

Figure 2.   TCE and IPA concentrations for a column operated in the down flow direction. Cumulative fluid injection commences at the time of initial IPA injection.

The column was initially saturated to 66%, and TCE saturation was reduced to 22% by waterflooding. Injection of a 1 PV slug of IPA reduced the TCE saturation to < 2%. The removal efficiencies of both the waterflooding and the alcohol-flooding processes were less efficient in this experiment because the column orientation minimized the favorable effects of gravitational forces. However, during the subsequent water flushing process, IPA was recovered from the column after only 3.5 PVs of total fluid injection. This process was more efficient than the down-flow experiment because the horizontal column orientation minimized the unfavorable effects of gravitational forces.

Results obtained from one-dimensional laboratory column experiments indicate that residual trichloroethylene globules left in a porous medium after waterflooding (pump-and-treat) may be removed by injecting isopropyl alcohol (2-propanol) through the contaminated zone. Subsequent water flushes through the treated zone can reduce the alcohol concentration to biodegradable levels. In our experiments, the balance of capillary, viscous, and gravitational forces were scaled to represent field conditions. Laboratory results were consistent with predictions from fluid flow theory for optimal flooding efficiencies. Additional work on this research, including the effects of alcohol concentrations and gravity-induced migration of DNAPL globules, is reported in Boyd and Farley (1992).

ACKNOWLEDGMENT

This research has been supported by the South Carolina Water Resources Research Institute, the Clemson University Research Office, and Sigma Xi, The Scientific Research Society.

REFERENCES

Boyd, G.R. and Farley, K.J., Residual NAPL removal from groundwater by alcohol flooding, *Trans Amer Geophys Union* 71(17):500, 1990.

Boyd, G.R. and Farley, K.J., NAPL removal from groundwater by alcohol flooding: Laboratory studies and applications, in *Hydrocarbon Contaminated Soils and Groundwater,* Volume 2, Kostecki, P.T. and Calabrese, E.J., (eds.), Lewis Publishers, Chelsea, MI, 1992.

Collins, R.E., *Flow of Fluids Through Porous Materials,* Reinhold Publishers Corporation, New York, 1961.

Gatlin, C. and Slobod, R.L., The alcohol slug process for increasing oil recovery, *Trans AIME* 219:46, 1960.

Kyte, J.R. and Rapoport, L.A., Linear waterflood behavior and end effects in water-wet porous media, *Trans AIME* 213:423-26, 1958.

Wilson, J.L., Conrad, S.H., Mason, W.R., Peplinski, W., and Hogan, E., Laboratory investigations of residual liquid organics from spills, leaks, and the disposal of hazardous wastes in groundwater, EPA/600/6-90/004, New Mexico Institute of Mining and Technology, Socorro, NM, 1990.

CHAPTER 31

Heterogeneities in Radionuclide Transport: Pore Size, Particle Size, and Sorption

Marilyn Buchholtz ten Brink, Susan Martin, Brian Viani, David K. Smith, and Douglas Phinney

Diffusive transport rates for aqueous species in porous media are a function of the molecular diffusion rate in solution for the species of interest, any change in its concentration in solution resulting from sorption or reaction, and the tortuosity of the material's pore structure. For heterogeneous natural materials and the complex solutions derived from many hazardous wastes, a mechanistic prediction of diffusion and transport rates is difficult without information about the relative importance of the various factors. An understanding of the effect of multiple transport paths and sorption mechanisms is particularly important for planning containment of hazardous radioactive waste because a small amount of radioisotope travelling via a faster than anticipated path may invalidate the predictions of transport models that assume homogeneous behavior.

For actinide elements and other hazardous materials, the coexistence of multiple oxidation states, colloidal species, and numerous complexes in ground or waste-water solutions suggests that a range of values exists for molecular diffusivity, reactivity, and sorptivity. Actinides tend to be very particle-reactive in that they adhere to existing surfaces and suspended particles or precipitate into colloid-sized complexes. Their transport rate can consequently be influenced by the natural colloid population and its behavior in situ. The physical structure of a rock often includes void spaces ranging in size from microscopic grain boundaries or intergranular pore spaces to fractures that are centimeters in width. This structure affects fluid-flow paths and colloid mobility, whereas heterogeneous mineralogy affects sorption.

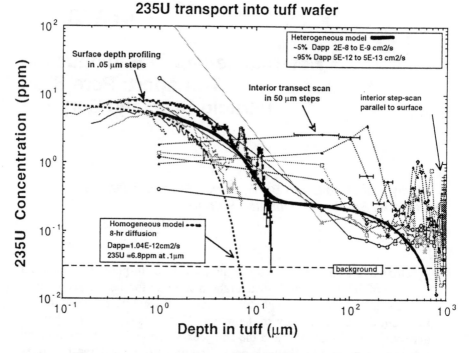

Figure 1. [235]U concentrations vs. depth obtained from surface depth profiles (lines with no symbols), interior step-scans perpendicular to the surface (lines with symbols), and interior step-scans parallel to the surface (crosses) on a wafer exposed to [235]U for 8 hours in a static-diffusion experiment. Different line patterns or symbols indicate separate analysis locations. Model fits are shown in bold lines.

In previous work, we measured diffusive transport of [235]U tracer in groundwater-saturated tuff rocks and found that regions of greater actinide identifiable microfractures, or concentrations of specific elements (Figures 1 and 2) (McKeegan et al. 1988; Buchholtz ten Brink et al. 1989). In addition, the abundance of colloid-sized particles and the apparent diffusivity increased when the sodium bicarbonate concentration was increased. To better understand these processes and to identify the impact of colloid size and composition on actinide mobility in the environment, we have undertaken to

1. characterize the physical structure of the porous media,
2. characterize naturally occurring colloids in areas of potential waste contamination,

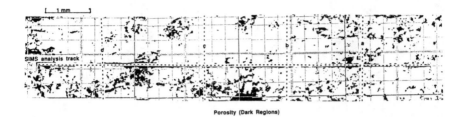

Porosity (Dark Regions)

Figure 2. Porosity in Topopah Tuff water interior. Darker regions indicate larger pores and represent the 3-5% of the total pore volume that has equivalent pore diameters of 10-100 μm. The line in the center is the SIMS analysis track and the dashed lines labelled with the letters a-d on the porosity image indicate locations of step scans taken perpendicular to the surface.

3. investigate the interaction of natural colloids with waste solutions,
4. characterize colloids generated from radioactive waste, and
5. model effects of changes in porous media and/or colloid properties on actinide mobility.

Results from (1) and (2) are presented here.

Rock samples were obtained from the Topopah Spring horizon of the Paintbrush Tuff (Figure 3a). The rock, a devitrified welded-tuff, having an average bulk porosity of 11%, was obtained from drill core (hole USW H-6). This rock is in contact with groundwater that flows to well J-13, Jackass Flats, Nevada. The pore size distribution was measured with mercury-intrusion porosimetry (Figure 4) and with image analysis of digitized scanning electron microscopy (SEM) photomicrographs (Figure 3b). Approximately 95 to 97% of the pores are <0.1 μm in equivalent diameter (mean 0.03 μm); the remaining 3 to 5% are between 10 to 100 μm (see Figure 2).

Rock samples from Pahute Mesa were obtained from four depths in hole Ue20n-1 (Figures 3b, 3c, and 4), which is located a few hundred meters from water well 20 at the Nevada Test Site (NTS) Nevada. These rocks, all rhyolitic welded tuffs, have a porosity of 13 to 15%. The pore size distribution was measured as it was for the Topopah Spring samples. The pore size distribution is unimodal with a mean between 1 to 3μm.

(a)   (b)   (c)

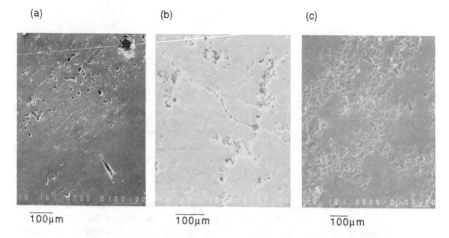

100μm   100μm   100μm

Figure 3. SEM photo-micrographs showing the >10 μm pores of (a) Topopah
Spring tuff wafer from 1109 ft. at 160x magnification and Pahute Mesa
tuff wafers from (b) 2415 ft and (c) 2633 ft. at 200x and 120x magnification (respectively).

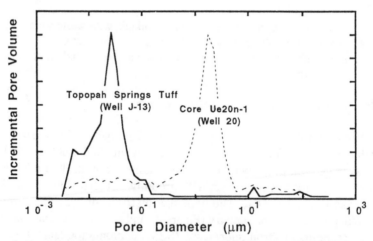

Figure 4. Pore-size distribution in Topopah Spring tuff (typical of 5 samples analyzed) and in Pahute Mesa tuff (typical of 4 samples) determined from
mercury-intrusion porosimetry.

Groundwater from well J-13 flows through the Topopah Spring tuff, and
water from well 20 flows through the Pahute Mesa tuff. Fluid and particulate phases were characterized in these waters. Samples of water were
collected in precleaned polyethylene bottles immediately after reaching the
surface. These wells are water supply wells operated at ~200 gal/min.

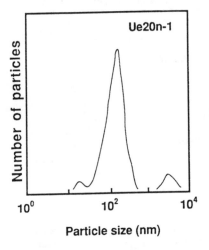
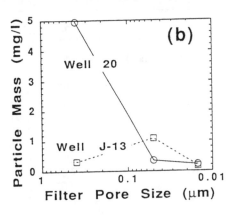

Figure 5. Size distribution of material (a) from well 20 concentrated by 10^3 MWCO ultrafilter and determined by autocorrelation photon spectroscopy and (b) from well 20 and well J-13 determined by sequential filtration with 0.4, 0.05, and 0.015 μm polycarbonate filters.

The bulk water properties of water from well J-13 and well 20 (Table 1) were determined using standard techniques.

The concentration of suspended particles and colloids was determined by filtration (Table 2), and the size distribution by autocorrelation photon spectroscopy (Figure 5a), sequential filtration (Figure 5b), and analysis of transmitted electron microscopy (TEM) photomicrographs. X-ray diffraction (XRD) and Fourier transform infrared spectroscopy (FTIR) were used to identify the mineralogy of material ≥0.015 μm collected on polycarbonate filters. Well 20 solids consisted of layer silicates, plagioclase, carbonate, and quartz. Positive identification of the layer silicate phases was not possible, but smectite, illite, and biotite are likely constituents.

The composition, size, and morphology of individual particles was assessed using TEM and energy dispersive spectroscopy (EDS) for solids in unfiltered water from well J-13. Particles include the following: 2 μm diameter Ca-Mg mineral; 1-μm Ca-Si oxide (with lesser Fe-Mg-Al); 1-μm Si oxide (Fe-Ca-Al-Mg); 0.5 and 1.0-μm Fe oxides (Si, Cr, Fe) with associated smaller particles (Figure 6a); 0.5-μm Al oxide aggregate (Figure 6b); 0.3-μm platey Al-Si oxide (Figure 6c); and 0.1-μm aggregate of Ca-Si-Ti (Fe, Ca, Al) oxide (Figure 6d). Particles observed in well 20 water (not shown) also encompass a broad range of sizes, compositions, and morphology. The association or coating of larger particles with smaller ones is

Table 1. Major chemistry of groundwater from well J-13 and well 20.

	J13	20		J13	20
Cations (mg/l)			Anions (mg/l)		
B	0.05	0.01	F	2.25	--
Ca	11.78	5.93	Cl	6.78	12.0
Fe	0.02	0.01	No_3	8.64	1.7
K	3.12	1.46	SO_4	18.22	31.4
Li	0.03	0.05	$HCO3$	125	111
Mg	1.85	0.22	TIC	25.7	20.3
Na	47.56	67.4	TOC	1.12	0.51
Si	30.03	23.0			
Sr	0.03	0.01			
Intake (ft)	1200	3200	Cond (μhmos)	--	320
Water table (ft)	929	2058	Eh (mv)	--	300
			pH	7.2	7.7
			Temp (°C)	32	40

Table 2. Concentration of suspended particles and colloids >0.015 μm in well waters.

Well	Mass (mg/l)	Samples
20	5.08	(1 2-l. sequential filtration)
J13	1.43	(1 2-l. sequential filtration)
J13	1.11±63	(7 30-ml filtrations)

common. Biological components were occasionally observed, and some particles seem to be tight aggregations of smaller particles.

Most of the pore volume of the rhyolitic tuff in contact with well 20 waters is in pores less than 5 μm in diameter, with a mean diameter of around 2 μm. This is larger than many of the particles identified in well20 water; consequently, we can expect fairly free movement of these colloids in this system. The bulk of the pore volume in the Topopah tuff in contact with J-13 water, however, is in pores <0.1 μm in diameter; thus, most of the colloids occurring in J-13 will have limited transport paths in the rock. A small fraction of pore volume in both rocks is attributed to larger pores or microfractures (10 to 100 μm) (Figure 4), which can be observed in 200X magnifications (Figure 3). These regions, and larger fractures that are identifiable in hand specimens, can provide a path for small but potentially significant transport of colloids and any material that becomes associated with them.

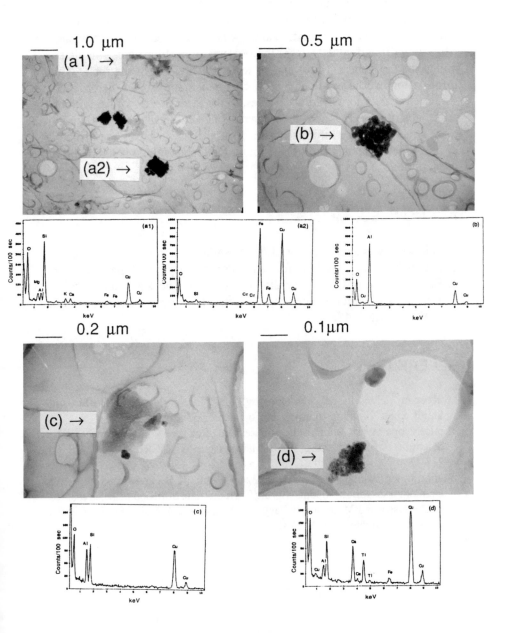

Figure 6. TEM and EDS analysis of solids in unfiltered water from well J-13. (a-d) Various magnifications showing particles from 1.0 to 0.1 μm in diameter and their chemical compositions. Cu in the EDS is from the TEM grid and the bubble-like features on the TEM photos are holes in the carbon film that coats the Cu grid.

In summary, 1) locations and sizes of "fast-transport" paths in porous media were identified, 2) the mass of particles available for transport or retention (a function of relative size) was determined for groundwaters from well 20 and well J-13, 3) the composition of particles was determined, and in general, most appear to be constituents of the rock. This information can be used to identify chemical stability and the probable sorptive capacity for hazardous wastes, 4) the identification of likely transport paths in these rocks for the suspended and colloidal material found in these waters provides information necessary to assess subsurface transport of hazardous materials and to plan any intentional alterations to colloid or waste mobility.

ACKNOWLEDGMENTS

This work was supported by LLNL Nevada Environmental and Applied Research Program. Work performed under the auspices of the U.S. Dept. of Energy Office of Civilian Waste Management, Yucca Mountain Project Office, by Lawrence Livermore National Laboratory under contract No. W-7405-ENG-48.

REFERENCES

McKeegan, K.D., Phinney, D., Oversby, V.M., Buchholtz ten Brink, M., and Smith, D.K., Uranium transport in Topopah Spring tuff: An ion-microscope investigation, Materials Research Society Symposia Proceedings: Scientific Basis for Nuclear Waste Management XII, UCRL 99734, 1988, pp 813-21.
Buchholtz ten Brink, M., Phinney, D., McKeegan, K.D., and Oversby, V.M., Uranium transport in Topopah Spring Tuff: multiple diffusion paths, Presented at Chemistry and Migration Behavior of Actinides and Fission Products in the Geosphere, Monterey, CA, 1989.

CHAPTER 32

Colloidal Borescope: A Means of Assessing Local Colloidal Flux and Groundwater Velocity in Porous Media

T.A. Cronk and P.M. Kearl

The colloidal borescope is a waterproof video camera capable of viewing indigenous colloids in a monitoring well. Through optical magnification, the movement and density of these colloids is easily assessed. The instrument shows promise in providing an improved methodology for determining both local groundwater flow velocity and colloidal transport potential.

Because observations are taken at the microscale, data obtained are indicative of the local transport parameters of the subsurface flow system. By taking numerous measurements at different spatial locations, heterogeneities of the flow system in both the vertical and horizontal sense may be defined. Preferential flow zones and fractures thus may be located. The capability to determine the spatial variability of relevant transport parameters is also necessary for the effective use of stochastic transport models.

Through determination of the functional relationships between advective colloidal transport in a borehole and specific discharge in the surrounding aquifer, local groundwater flow velocity (both magnitude and direction) is also determinable. Experimental observations to date indicate a steady, laminar flow field in the borehole that has an excellent directional correlation with specific discharge in the surrounding aquifer. Conversely, the experimentally observed relationships between wellbore velocity and specific discharge is not in agreement with the predictions of potential flow theory. Although borehole flow velocities are in relative agreement with specific discharge (i.e., high flow zones exhibit high borehole velocities and low flow zones exhibit low borehole velocities), these same measurements are consistently higher than values predicted by potential flow theory. The observed relationships suggest an exciting new means of investigating the physical phenomena affecting flow in porous media. These observations offer a means of enhancing conceptual understanding and of developing improved transport theory models.

Use of the borescope to confirm the effective radius of influence of groundwater extraction systems is of value in determining the efficiency of groundwater remediation systems. This is accomplished by observing groundwater flow in an observation well placed on the design perimeter of an extraction system to determine if flow is actually in the direction necessary for extraction. This measurement capability would be invaluable in demonstrating regulator compliance.

Field use of the colloidal borescope has indicated that colloidal density and migration velocity are parameters that are very sensitive to external perturbations. The slightest disturbance induces pressure waves and turbulent flow into the wellbore, causing changes in these parameters over several orders of magnitude. An interesting field experiment confirming this sensitivity involved dropping a slug into a well some 5 m away from a well in which the instrument was recording steady-state movement. The results were dramatic as the colloidal migration pattern literally exploded into a turbulent flow pattern. One can imagine the affects even the most gentle pumping techniques must have on turbidity and colloid density. These observations indicate the instrument may be invaluable in determining when steady, undisturbed flow conditions exist and hence providing a much more accurate measurement of indigenous colloidal densities and migration rates. This information would also be of value in assessing the need for filtration of groundwater samples for chemical analysis.

The instrument's observational capabilities may also prove useful as an investigative tool for improving theoretical conceptualization of flow processes at the most basic level. As mentioned, experimental observations have recorded consistently higher velocity magnitudes than those predicted by potential flow theory. This indicates that the continuum potential flow model may be inappropriate for modeling flow at the scale of observation. The development of a consistent theoretical explanation for these experimentally observed relationships is sure to provide exciting new knowledge about the underlying physical processes.

CHAPTER 33

Uptake of Hydrophobic Organic Compounds from Soil by Colloidal Surfactant Micelles

David A. Edwards, Zhongbao Liu,
and Richard G. Luthy

Surfactants are a class of amphiphilic compounds distinguished from other compounds by their surface-active properties. An important property of most surfactants is their ability at sufficiently high concentrations in aqueous solution to spontaneously form colloidal aggregates known as micelles. The hydrophobic moieties of the surfactant molecules of a micelle are oriented toward its interior; the hydrophilic moieties of the molecules face outward toward the surrounding solution. Although a micellar surfactant solution is rigorously defined as consisting of a single, isotropic, thermodynamically stable phase (Attwood and Florence 1983), it is often advantageous to consider such a solution as consisting of two pseudophases: a micellar pseudophase, comprising all of the hydrocarbon-like interior portions of the surfactant micelles, and an aqueous pseudophase, consisting of the surrounding hydrophilic solution saturated with surfactant monomers Valsaraj and Thibodeaux 1989). In addition, because the micellar pseudophase is hydrophobic, it is capable of solubilizing solutes having limited solubility in hydrophilic liquids [e.g., hydrophobic organic compounds (HOCs)].

Remediation technologies based on in situ surfactant solubilization of HOCs may at some future time permit the desorption, mobilization, and recovery of otherwise highly sorbed HOC from humic and mineral surfaces in the subsurface. Several engineering problems are currently associated with implementing such a technology, however, and our understanding of the basic physicochemical processes involved is limited.

As part of an ongoing investigation of surfactant solubilization processes, laboratory experiments (Edwards et al. 1991) and modeling (Edwards et al. 1990) dealing with nonionic surfactant solubilization of HOCs in both aqueous and soil/aqueous systems have been conducted. The HOCs studied are the polycyclic aromatic hydrocarbons (PAHs) naphthalene, phenanthrene, and pyrene. The surfactants studied comprise an alkylethoxylate (C_4E_{12}), which, like other alkylethoxylates, is relatively nontoxic, and three alkyl-phenol ethoxylates, $C_8PE_{9.5}$, C_8PE_{12}, and $C_9PE_{10.5}$, which, while useful for the current study, are probably not viable candidate surfactants for field application.

Solubilization of HOC is initiated when micelles start forming at a critical micelle concentration (CMC). The amount of HOC solubilized in micellar solutions is typically a linear function of the concentration of the surfactant in micelle form. The distribution of HOC between micellar and aqueous pseudophases can be quantified with a mole fraction micelle/water partition coefficient K_m, which represents the ratio of the mole fraction of the HOC in the micellar pseudophase X_m, to the mole fraction in the aqueous pseudophase, X_a (Hayase and Hayano 1977).

$$K_m = X_m/X_a \tag{1}$$

Batch test solubilization of PAH was conducted in 8-mL vials with duplicate samples of 5-mL micellar solution for each data point. Measured amounts of surfactant at various concentrations and a uniform quantity of PAH stock with 20 to 80 times the PAH mass needed to attain aqueous solubility were added to the set of samples representing a given surfactant-PAH combination. The PAH stock comprised both non labeled and ^{14}C-labeled PAH of predetermined mass ratio. The activities of the ^{14}C-labeled PAH solutions were counted using a Beckman LS 500 TD Liquid Scintillation Counter (LSC). The sample vials were sealed with Teflon-lined septa and open-port screw caps and set in a reciprocating water bath, which was cycled at 80 cycles per minute for approximately 24 h at 25 C. Duplicate aliquots were taken from each sample by glass syringe and filtered through 0.22-µm-diam pores in pre-equilibrated Teflon membranes to block passage of crystalline PAH. Aliquots of 0.5-to 2.0-mL, depending on the series of experiments, were injected into 10-mL liquid scintillation cocktail (Scintiverse II, Fisher Scientific). The decay rates of the ^{14}C-labeled material in the aliquots were measured with the H# quench monitoring and automatic quench compensation technique in the LSC. Background decay rate corrections were made and calculations were performed to determine PAH apparent solubilities as a function of surfactant dose.

Other tests for determining the CMC and the amount of sorption for each surfactant in aqueous and soil/aqueous systems, respectively, employed measurement of the surface tension of the bulk solution as a function of the logarithm of the surfactant dose applied.

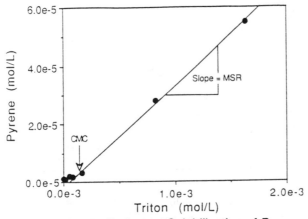

Figure 1. Triton X-100 Surfactant Solubilization of Pyrene.

Solubilization for each of the 12 surfactant-PAH combinations in aqueous systems was initiated at the CMC, as confirmed by separate surface tension experiments. Apparent PAH solubilities were linearly proportional to micellar surfactant concentration, as illustrated in Figure 1. The molar solubilization ratio (MSR), defined as the number of moles of HOC solubilized per mole of micellar surfactant, was determined for each combination, and a value of K_m was calculated using Eq. 2:

$$K_m = [(55.4 \ / \ C_{aq}) \ (MSR \ / \ (1+MSR)] \qquad (2)$$

where C_{aq} denotes the PAH concentration in the aqueous pseudophase. In the presence of pure-phase PAH in the aqueous systems, C_{aq} was a constant, equal to S_{cmc}, the apparent PAH solubility at the CMC; the MSR and K_m also had constant values. In the soil/water systems in the absence of PAH as a separate phase, both C_{aq} and the MSR varied as a function of surfactant dose, whereas the value of K_m was invariant. Log K_m values for various surfactant-PAH combinations ranged from 4.57 to 6.53.

For the aqueous-phase CMC to be attained, the quantity of nonionic surfactant that must be added to an aqueous system is much less than that which must be added to a soil/water system of the same aqueous volume because of surfactant sorption onto soil. Sorption of nonionic surfactant onto soil also modifies the HOC sorption coefficient as a result of two countervailing factors: an increase in the organic carbon content of the soil and an increase in the HOC apparent solubility in the aqueous pseudophase. The net result can be estimated for systems of soil and micellar surfactant solution with the HOC sorption parameter $K_{d,cmc}$:

$$K_{d,cmc} = K_d(S \ / \ S_{cmc})(f'_{oc} \ / \ f_{oc}) = K_{oc}(S \ / \ S_{cmc})(f'_{oc}) \qquad (3)$$

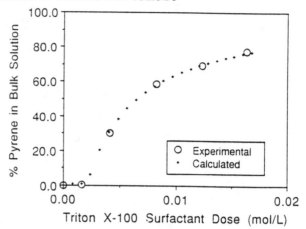

Figure 2. % Pyrene in Bulk Solution as a Function of Triton X-100 Dose.

where K_d and K_{oc} are linear HOC soil/water partition coefficients in the absence of surfactant, the latter of which is normalized for soil organic carbon content in nominal units of liters per gram; S and S_{cmc} are the solubilities of the HOC in moles per liter in pure water and in surfactant solution at the CMC, respectively; and f_{oc} and f'_{oc} are the fractional organic carbon contents of the soil, before and after the sorption of surfactant, respectively.

Micellar surfactant solubilization of HOC and sorption of HOC in systems of soil and micellar surfactant solution can be modeled using parameters and variables obtained from independent tests. F, the fraction of HOC in bulk solution, is given by

$$F = v_a + [(K_m V_w)(m_{tot} - Q_{surf} w_{soil} - v_a CMC)/(1 - K_m V_w C_{aq})]/$$
$$v_a + [(K_m V_w)(m_{tot} - Q_{surf} w_{soil} - v_a CMC)/(1 - K_m V_w C_{aq})] + w_{soil} K_{d,cmc} \quad (4)$$

where C_{aq} is a defined function of K_m, V_w, $K_{d,cmc}$, w_{soil}, v_a, C_{init}, m_{tot}, and Q_{surf}. C_{init} represents the original aqueous-phase organic compound concentration before the addition of surfactant to the soil/aqueous-phase system v_a is the volume of the bulk aqueous phase V_w is the molar volume of water (0.01805 mole/L at 25 C) m_{tot} is the total number of moles of surfactant added to the system Q_{surf} is the number of moles of surfactant sorbed per gram of soil and w_{soil} is the weight of the soil in grams. With values of C_{aq}, v_a, and the total number of moles of HOC present in the system, Eq. 4 permits the calculation of HOC concentrations in the aqueous pseudophase, the micellar pseudophase, and the sorbed phase.

Figure 2 shows the results of using Eq. 4 to model $C_8 PE_{9.5}$ surfactant solubilization of pyrene in a system of 0.045 L of water and 6.25 g of soil. The model employs variable and parameter values obtained from independent experimental tests. Calculated values of percentage pyrene in bulk solution (F x 100%) are plotted as a function of surfactant dose and compared with the experimental data of Liu et al. The close agreement exemplifies the potential usefulness of the model for predicting HOC partitioning in soil/aqueous systems of uniform surfactant concentration.

ACKNOWLEDGMENT

This work was sponsored by the U.S. Environmental Protection Agency Office of Exploratory Research under grant number R-816113-01-0.

REFERENCES

Attwood, D. and Florence, A.T., *Surfactant Systems: Their Chemistry, Pharmacy and Biology*, Chapman and Hall, New York, 1983.

Edwards, D.A., Luthy, R.G., and Liu, Z., Solubilization of polycyclic aromatic hydrocarbons in micellar nonionic surfactant solutions, (in press).

Edwards, D.A., Liu, Z., and Luthy, R.G., Physicochemical characterization of hydrophobic organic compound distribution in systems of nonionic surfactant solution and soil, Manuscript in preparation, Carnegie Mellon University, (1990).

Hayase, K. and Hayano, S., The distribution of higher alcohols in aqueous micellar solutions, *Bull. Chem. Soc. Japan* 50(1):83-5, 1977.

Liu, Z., Laha, S., and Luthy, R.G., Surfactant solubilization of polycyclic aromatic hydrocarbons in soil/water suspensions, *Water Sci Tech* 23:475-85, 1991.

Valsaraj, K.T. and Thibodeaux, L.J., Relationship between micelle-water and octanol-water partition constants for hydrophobic organics of environmental interest, *Water Res* 23(2):183-89, 1989.

CHAPTER 34

Deposition of Colloidal Particles in Porous Media in the Presence of Attractive Double Layer Interactions

Menachem Elimelech

The capture of colloidal particles from flowing suspensions by stationary surfaces (commonly referred to as *deposition*) is of direct practical concern in many technological and natural processes encountered in environmental science and engineering. Examples are transport of viruses, bacteria, and colloid-associated pollutants in subsurface flow (McDowell-Boyer et al. 1986), granular filtration in water and wastewater treatment (Yao et al. 1971; Tien and Payatakes 1979), and deposition of bacterial cells during biofilm formation (Bryers and Characklis 1982). Deposition is also of fundamental interest as a means of investigating the various colloidal and hydrodynamic forces involved in particle-particle interaction phenomena (Elimelech and O'Melia 1990; Gregory and Wishart 1980; Tobiason 1989).

Deposition of colloidal particles in porous media in the presence of attractive double-layer interactions has not yet been studied either theoretically or experimentally. A systematic investigation is presented in this study. Deposition experiments of positively charged model colloids on negatively charged collectors packed in a laboratory-scale column were conducted. The results were compared with modern theories of mass transfer, physicochemical hydrodynamics, and colloidal interactions. The role of colloidal and hydrodynamic interactions in the kinetics of colloid deposition in the presence of attractive double-layer interactions was evaluated and discussed.

Results of the deposition experiments at various KCl concentrations are presented as particle breakthrough curves in Figure 1. The physical conditions (i.e., collector size, media depth, and approach velocity) in those deposition experiments were similar. As a result, changes in deposition

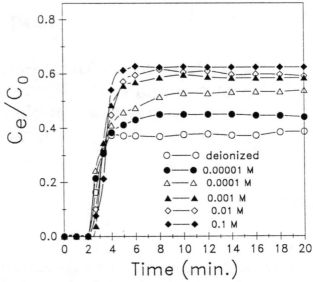

Figure 1. Particle breakthrough curves at different KCl concentrations. In all experiments the approach velocity was 1.4 mm/s, the collector size was 0.4 mm, the particle size was 0.477 μm, and the media depth was 160 mm.

with varying electrolyte concentrations can be attributed mainly to the effect of double-layer interactions. The results presented in Figure 1 show that removal of particles by the packed bed increases (C_e/C_0 decreases) with a decrease in the ionic strength of the solution. Removal of particles is less sensitive to ionic strength at electrolyte concentrations above 10^{-3} M. Once a complete breakthrough of particles is attained (at ~10 min), the removal rates of particles at each electrolyte concentration are constant. The results described in Figure 1 suggest that the influence of double-layer attraction forces increases with a decrease in the ionic strength of the solution, causing an increase in particle removal rates.

To delineate the effect of colloidal interactions on deposition rates under attractive double-layer interactions, it is imperative that experimental particle deposition rates be calculated from the breakthrough curves. Particle deposition rates in the early stage of the deposition experiments can be calculated from the initial removal of particles (C_e/C_0) obtained from the particle breakthrough curves (Yao et al. 1971; Elimelech and O'Melia 1990).

Ratios of the experimental deposition rate η_{exp} to the theoretical deposition rate calculated for pure convection diffusion η_0 (i.e., without colloidal and hydrodynamic interactions) are presented as a function of ionic

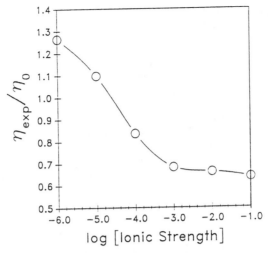

Figure 2. The ratio η_{exp}/η_0 as a function of the ionic strength (in M) of the solution.

strength in Figure 2. η_0 can be obtained from the classical solution of Levich (Yao et al. 1971; Levich 1962) as follows:

$$\eta_0 = 4.0A_s^{\frac{1}{3}}\left(\frac{D_\infty}{Ud_c}\right)^{\frac{2}{3}} \tag{1}$$

Where A_s is a porosity-dependent flow parameter to account for the effect of neighboring collectors on the flow field based on Happel's (Happel 1958) model; U is the approach velocity of the fluid; d_c is the collector diameter; and D_∞ is the diffusion coefficient at infinite separation, calculated from the Stokes-Einstein equation. The deposition rate described by Eq. (1) is obtained from the so-called Smoluchowski-Levich approximation. In this approximation, the transport equation without colloidal and hydrodynamic interactions is solved for the deposition rates; it is assumed that hydrodynamic interaction (retardation) and van der Waals attraction counterbalance each other. As observed from the results shown in Figure 2, deposition rates increase by almost a factor of 2 when the ionic strength is decreased from 0.1 to 10^{-6} M. The ionic strength of the deionized water at pH ~6 was assumed to be 10^{-6} M (as the concentration of the H^+). The experimental deposition rates at ionic strengths of 10^{-5} and 10^{-6} M are larger than those predicted for pure convection diffusion [Eq. (1)]. At

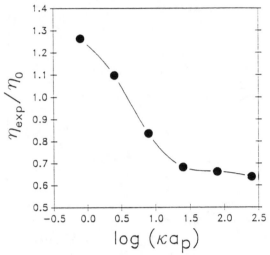

Figure 3.   The dimensionless ratio η_{exp}/η_0 as a function of the logarithm of the dimensionless parameter κa_p.

ionic strengths lower than 10^{-4} M, the deposition rates are smaller than those for pure convection diffusion.

Numerical solution of the transport equation reveals that the dimensionless deposition rate of particles in the presence of colloidal interactions can be written as a function of several dimensionless parameters (Prieve and Lin 1980).

$$\eta = \eta\,(N_{Pe}\,,\ N_V\,,\ N_E\,,\ ka_p)\qquad\qquad(2)$$

Here N_{Pe} is a Peclet number defined as Ua_p/D_∞, where a_p is the particle radius. The van der Waals interactions are characterized by $N_V = A/kT$, where A is the Hamaker constant of the interacting media, k is the Boltzmann constant, and T is the absolute temperature. Electrical double-layer interactions are characterized by $N_E = \varepsilon_0\varepsilon_r\psi_1\psi_2 a_p/kT$, where ε_0 and ε_r are the permittivity in vacuum and the relative permittivity of water, respectively, and ψ_1 and ψ_2 are the surface potentials of particles and collectors, respectively. The parameter κa_p characterizes the thickness of the diffuse layer; κ is the inverse Debye length. It is controlled by the ionic strength of the solution.

The ratio of the experimental deposition rate to the deposition rate for pure convection diffusion (η_{exp}/η_0) as a function of the logarithm of the diffuse layer thickness parameter (κa_p) is presented in Figure 3. It is demonstrated that deposition rates increase significantly with an increase in

κa_p. The increase in the deposition rates is the result of an increase in the range of the attractive double-layer forces as the ionic strength decreases. This finding is in accord with theoretical predictions of Chari and Rajagopalan (1985) and Adamczyk et al. (1989) for particle deposition in a stagnation point flow in the presence of attractive double-layer interactions. Their numerical solution of the transport equation in the presence of attractive double-layer interactions indicates that the parameter κa_p has a marked effect on particle deposition rates. Theoretical predictions for particle deposition in a stagnation point flow in the presence of attractive double layer interactions (Chari and Rajagopalan 1985; Adamczyk et al. 1989) demonstrated that the increase in the range of the double-layer forces resulting from a decrease in ionic strength makes a much larger contribution to the enhancement of particle deposition rates than does the increase in ζ potentials (through the parameter N_E).

This inverse electrolyte effect is opposite to that commonly observed in deposition and coagulation kinetics of similarly charged surfaces, where particle deposition or coagulation rates increase with an increase of ionic strength. The findings of this study contradict the commonly accepted supposition that deposition rates of oppositely charged colloids and surfaces are comparable to barrierless deposition rates of similarly charged particles and collectors (i.e., when energy barriers in the total interaction energy profile are negligible).

Increased deposition rates of oppositely charged colloids and surfaces at low ionic strengths are attributed to an increase in the range and magnitude of the double layer attraction forces. Transport of particles as a result of long-range double-layer attraction energies can increase beyond that caused by convection and diffusion. As the ionic strength increases, the range and magnitude of the double-layer attraction energies decrease and, as a result, deposition rates decrease. The overall kinetics of colloid deposition in the presence of attractive double-layer interactions are determined by a balance between attractive forces, originating from van der Waals and double-layer interactions, and repulsive forces, resulting from hydrodynamic interaction (retardation).

REFERENCES

Adamczyk, Z., Siwek, M., Zembala, M., and Warszynski, P., *J Colloid Interface Sci* 130:578, 1989.
Bryers, J.D. and Characklis, W.G., *Biotechnol Bioeng* 24:2451, 1982.
Chari, K. and Rajagopalan, R., *J Chem Soc Faraday Trans* 2:81:1345, 1985.
Elimelech, M. and O'Melia, C.R., *Langmuir* 6:1153, 1990.
Gregory, J. and Wishart, A.J., *Colloids Surf* 1:313, 1980.
Happel, J., *AIChE J* 4:197, 1958.

Levich, V.G., *Physicochemical Hydrodynamics*, Prentice Hall, NJ, 1962.

McDowell-Boyer, L.M., Hunt, J.R., and Sitar, N., *Water Resour Res* 22:1901, 1986.

Prieve, D.C. and Lin, M.M.J., *J Colloid Interface Sci* 76:32, 1980.

Tien, C. and Payatakes, A.C., *AIChE J* 25:737, 1979.

Tobiason, J.E., *Colloids Surf* 39:53, 1989.

Yao, K.M., Habibian, M.T., and O'Melia, C.R., *Environ Sci Technol* 5:1105, 1971.

CHAPTER 35

Sizing and Chemical Characterization Studies of Natural Colloids and Macromolecules

J.S. Gaffney, N.A. Marley, and K.A. Orlandini

Recent work has indicated that submicron colloidal and macromolecular material may act as an important means of transporting hydrophobic and hydrophilic contaminants in groundwaters (McCarthy and Zachara 1989). Indeed, the movement of actinides in subsurface environments has been shown to be strongly correlated with small colloids and macromolecular fractions within the groundwater (Penrose et al. 1990). If we are to understand the interactions of subsurface contaminants with natural materials in subsurface environments, we will need to develop methods for physical and chemical characterization that will allow us to evaluate the importance of natural humic and fulvic materials in these complex systems. Recently, we have been evaluating the use of hollow-fiber ultrafilters for the size fractionation of colloidal and macromolecules in water after the removal of large particulates and microorganisms by use of a 0.45-μm filter (Orlandini et al. 1990; Marley et al. unpublished data). When the samples have been sized with the filters (Amicon), we use a variety of analytical techniques to characterize the inorganic and organic (humic and fulvic acids) components in the samples (Marley et al. unpublished data).

Physical sizing is carried out by using 0.1 μm and 100, 30, 10, and 3K (mol. wt × 10^{-3}) hollow-fiber ultrafilters. After filtration through the 3K filter, membrane filters are used to evaluate the sample at 1 and 0.5K sizes. We have been able to obtain enough of the sized material to perform chemical characterization on a number of surface and subsurface waters by using pyrolysis/gas chromatography, infrared spectroscopy, and actinide and trace metal analyses.

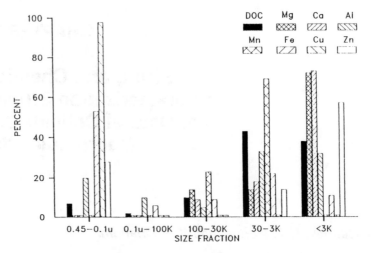

Figure 1.   Dissolved organic carbon (DOC) and trace metal size distribution from a representative surface water.

Typical results to date indicate that inorganic trace metals and actinides (thorium, uranium, plutonium, and americium) are not evenly distributed as a function of size of the associated particles. A typical example of the dissolved organic carbon (DOC) and trace metal distribution observed in a surface water sample is given in Figure 1. For surface water samples we typically observe a bimodal distribution with appreciable metal and actinide content in the 0.1- to 0.45-μm size range and the 500 to 10K molecular weight size range. The former group appears to be organically coated inorganic particles, while the latter has been shown by spectroscopic methods to be principally fulvic acids (Marley unpublished data). Indeed, we have observed that the thorium isotopes 228, 230, 232, and 234 are not evenly distributed in many water samples; increased levels of isotope 228 are particularly apparent in the small size ranges (Gaffney unpublished data). Because thorium isotopes are expected to have identical chemistries and the thorium-228 isotope has a half-life of approximately 2 years, these data suggest that once bound to these active complexing agents, the thorium isotopes do not readily equilibrate in the water. These observations for natural actinide systems are consistent with the data observed for contaminants (e.g., plutonium and americium; N. Marley, Argonne National Laboratory, unpublished data).

Subsurface waters have been observed to be typically unimodal distributions, with the inorganic species being complexed in the smaller fractions between 0.5 and 30K. Chelating agents typically in the size fraction 0.5 to 3K, have been shown to actively bind metals and actinides. An example of the DOC and trace metal distributions observed with hollow-

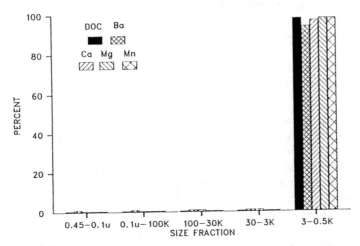

Figure 2.   DOC and trace metal size distribution for a representative groundwater.

fiber ultrafiltration size fractionation of the samples is shown in Figure 2.
With diffuse reflectance infrared Fourier transform (DRIFT) spectroscopy,
these organic fractions have been characterized as humic and fulvic acids.
These materials are actually organometallic in nature because they typically contain a significant weight percentage of metals. In a number
of samples, spectroscopic evidence for covalently bound silica-organic
esters in the humic and fulvic materials was obtained. This observation is
consistent with previous work on model carboxylate-silica interactions
(Marley et al. 1989; N. Marley, Argonne National Laboratory, unpublished
data).
     The results to date indicate that the active aqueous chelating materials
are small macromolecules, which can be powerful agents in binding and
transporting metals. The size fractionated samples also clearly indicate
that many inorganic metals and actinides are not in equilibria. Thus, use
of equilibrium based models to predict the chemical and physical distribution of toxic metals or radionuclides (e.g., Pu, Am, and Np) will probably
lead to inaccurate prediction of the transport of these materials in the subsurface. These observations are consistent with field work at Mortandad
Canyon, New Mexico (Penrose et al. 1990). Models had predicted that
plutonium and americium would exist in chemical forms that would not be
transported but would be confined in the volcanic tuff matrix of the aquifer. Instead, these species were observed as far as 3000 m from the initial
outfall (Penrose et al. 1990). Thus, a thorough understanding and chemical
characterization of these colloidal and macromolecular complexors will aid

in prediction of contaminant transport; in cleanup of already existing waste sites; and in barrier design for future handling of organic, inorganic, radioactive, and mixed wastes.

REFERENCES

Marley, N.A., Bennett, P., Janecky, D.R., and Gaffney, J.S., Spectroscopic evidence for organic diacid complexation with dissolved silica in aqueous systems-I, Oxalic acid, *Org Geochem* 14:525-28, 1989.

McCarthy, J.F. and Zachara, J.M., Subsurface transport of contaminants: Mobile colloids in the subsurface environment may alter the transport of contaminants, *Environ Sci Technol* 23:496-502, 1989.

Orlandini, K.A., Penrose, W.R., Harvey, B.R., Lovett, M.B., and Findlay M.W., Colloidal behavior of actinides in an oligotrophic lake, *Environ Sci Technol* 24:706-12, 1990.

Penrose, W.R., Polzer, W.L., Essington, E.H., Nelson, D.M., and Orlandini, K.A., Mobility of plutonium and americium through a shallow aquifer in a semiarid region, *Environ Sci Technol* 24:228-34, 1990.

CHAPTER 36

Characteristics of Colloids in Landfill Leachate

V. Gounaris, P.R. Anderson, and T.M. Holsen

Leachate was collected from a municipal landfill that has been closed for about 15 years. The sample was collected (under a N_2 atmosphere) only after observations of pH, specific conductance, turbidity, and D.O. suggested that the sample was representative of the in situ leachate. Ultrafiltration membranes were used to separate the bulk leachate into several size fractions. Preliminary results indicate that about 40% of the TOC and 50% of the total Fe are in the colloidal size fraction.

Landfilling has been a major disposal method for industrial and domestic wastes for many years. Despite its popularity as a treatment option, there is an incomplete understanding of the chemical and physical characteristics of leachates that can be generated from landfills. This is especially true of the colloidal fraction, which can have an effect on the fate or subsequent processing of leachate. For example, in investigations of leachate-contaminated groundwater, solute transport models that ignore colloid--facilitated transport may seriously underestimate contaminant mobility (Gschwend and Reynolds 1987). Furthermore, the presence of colloids in landfill leachate can complicate treatment strategies (Smith and Weber 1988; Gourdon et al. 1989). The overall objective of this research project is to describe the characteristics of colloidal materials in the leachate from municipal landfills and determine which of the colloidal size fractions contaminants are associated with. Ultrafiltration will be used to segregate specific size fractions of colloids to determine variations in their properties as a function of particle size. Specific objectives within the overall objectives include:

- Determining physical and chemical characteristics of municipal landfill leachate.

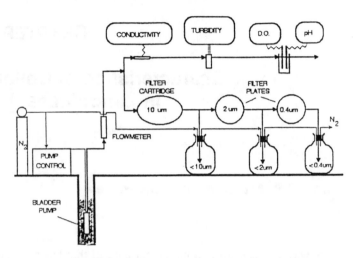

Figure 1. Sampling apparatus for field collection of leachate.

- Separating discrete colloid fractions from the bulk leachate by ultrafiltration.
- Determining physical and chemical characteristics of the segregated fractions.
- Identifying correlations that may exist between representative contaminants and the colloid fractions.
- Evaluating surface chemical properties of the colloidal materials to estimate their transport potential.

Leachate samples were collected from a closed landfill that primarily received mixed municipal wastes during the 1960s and 1970s. Samples were collected from a depth of 8 m by using a N_2-driven, Teflon bladder pump and Teflon tubing at a flow rate of 100 mL/min (Figure 1). Prior to sample collection, the leachate was routed through a monitoring train to observe changes in specific conductance, turbidity, D.O., and pH. After 2 hours of pumping the leachate appeared to be in steady-state condition (Figure 2), and a sample was collected. Three size fractions were collected in the field: < 0.4 μm, 0.4 to 2 μm, and 2 to 10 μm. The smallest size fraction was further fractionated in the laboratory.

These results are preliminary. Laboratory tests with model colloids demonstrated that polycarbonate membrane ultrafilters have a more well-defined pore size and minimal sorption losses relative to polysulfone membranes. Some of the information summarized here was developed with polysulfone membranes. The continuing work is using membranes in the following size ranges: 10 to 1.0 μm, 1.0 to 0.1 μm, 0.1 to 0.01 μm, and 0.01 to 0.001 μm. Only the smallest of these membranes will be poly-

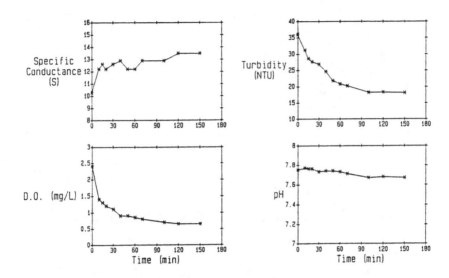

Figure 2. Assessment of steady state conditions during leachate collection.

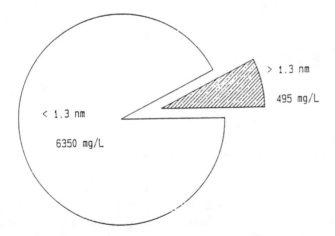

Figure 3. Total solids (10 μm) distribution in the leachate.

sulfone. Because the greatest sorption losses occur on the membrane support material rather than on the membrane surface, sorption losses should be minimal.

The size distribution of total solids in the bulk leachate (<10 μm) indicates that about 7% of the solids are in the colloidal size range (Figure 3). Currently, a more-detailed description of the colloids in the leachate is

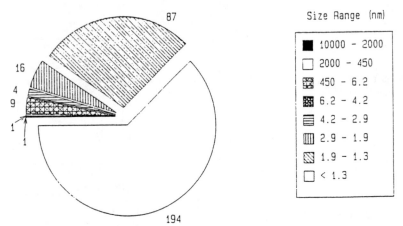

Figure 4. Total organic carbon (<10 µm) distribution in the leachate.

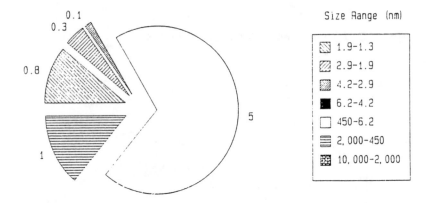

Figure 5. Colloidal iron distribution in the leachate.

limited to the organic carbon and iron content. Almost 40% of the total organic carbon (TOC) in the bulk leachate (<10 µm) is in the colloidal size range (Figure 4). This is a significant amount of organic matter (118 mg/L), which may serve as a partitioning medium for hydrophobic contaminants in the leachate. Information about the inorganic colloidal material in the leachate is currently limited to iron, and about 50% of the total iron (≈ 10 mg/L) was found in the colloidal size range. Interestingly, most of this material was in the range from ~6.2 to 450 nm rather than in the smallest size faction.

Samples of the bulk leachate will undergo PIXE (Proton Induced X-ray Emission) analysis for elemental composition and gas chromatography/mass spectroscopy (GC/MS) analysis for base/neutral extractable (EPA

Table 1. Summary of the organic semivolatile content of the leachate

Compound	µg/L
1,4-Dichlorobenzene	16.3
1,2-Dichlorobenzene	3.2
2,4-Dimethylphenol	12.5
Naphthalene	9.3
Dimethyl phthalate	31.0
Di-N-Butyl phthalate	14.3
Bis(2-Ethylhexyl) phthalate	17.2

Method 625) and poly chlorinated biphenyls and pesticides (EPA Method 608). Based on these results, we hope to identify approximately 12 representative compounds that we will try to track through the various size fractions. The initial results from the GC/MS analysis (Table 1) have identified seven likely candidates. Finally, potentiometric titrations and observations of electrophoretic mobility will be used to describe surface chemical properties of some of the colloids.

ACKNOWLEDGMENTS

This research was made possible by a grant from the Center for Solid Waste Management and Research, University of Illinois. We would also like to thank the landfill operators for permission to collect samples on site.

REFERENCES

Gourdon, R., Comel, C., Vermande, P., and Veron, J., Fractionation of the organic matter of a landfill before and after aerobic or anaerobic biological treatment, *Water Res* 23:167-73, 1989.

Gschwend, P.M. and Reynolds, M.D., Monodisperse ferrous phosphate colloids in an anoxic groundwater plume, *Contam Hydrol* 1:309-27, 1987.

Smith, E.H. and Weber, W.J., Modeling activated carbon adsorption of target organic compounds from leachate-contaminated groundwaters, *Environ Sci Technol* 22:313-21, 1988.

CHAPTER 37

Colloidal Radionuclide Groundwater-Surface Water Relationships

Scott R. Grace

An initial study was conducted to determine the size of plutonium and americium in surface water seeps at the Rocky Flats Superfund site near Golden Colorado (Figure 1). Historical data show plutonium and americium in the "dissolved" category (less than 0.45 μm). Due to significant surface soils contamination and the nature of the seeps (intermittent flow), the high suspended sediment load, and the typically "hydrophobic" nature of plutonium and americium, there is uncertainty as to whether the filtered samples contained truly dissolved species or simply colloidal complexes in the groundwater and/or surface water seeps. Preliminary data indicate that plutonium concentrations are primarily greater than 0.45 μm but americium concentrations are significant below 0.2 and 0.1 μm, indicating the potential presence of colloidal complexes. Additional studies are planned to further determine whether the colloidal complexes are particulate, organically bound, and/or clay-mineral.

The source of the plutonium and americium was the 903 Pad drum storage area that was used from 1958 to 1967. The drums, some of which corroded and leaked, contained oils and solvents contaminated with plutonium/americium. Most of the drums contained lathe coolant (mineral oil and carbon tetrachloride in varying proportions), but some also contained hydraulic oils, vacuum pump oils, trichloroethene, tetrachloroethene, silicon oils, acetone, and ethanolamine. An estimated 5000 gal. of liquid (containing approximately 86 g of plutonium and 14 g of americium) leaked into the soil during use of the drum storage area. During drum removal and attempted remediation activities in the late sixties, winds distributed plutonium/americium beyond the pad area. Plutonium is significantly elevated (up to 500 pCi/g) in surface soils.

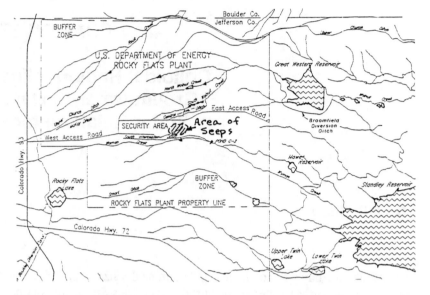

Figure 1. Location map.

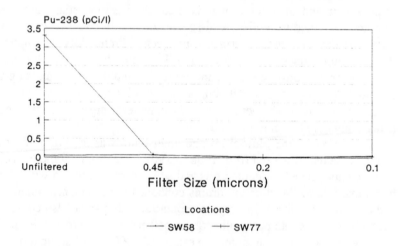

Figure 2. Plutonium-238 Concentration (pCi/l).

Nuclepore polycarbonate filters, 0.45, 0.2, and 0.1 μm, were used in an initial investigation conducted June 1990. Unfiltered water, filtered water, and filters were analyzed for ^{238}Pu ^{239,240}Pu, and ^{241}Am. Results of the

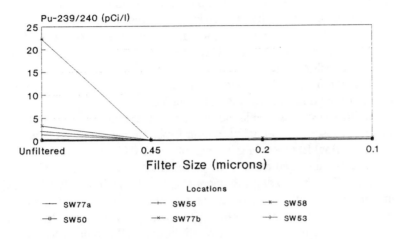

Figure 3. Plutonium-239/240 Concentration (pCi/l).

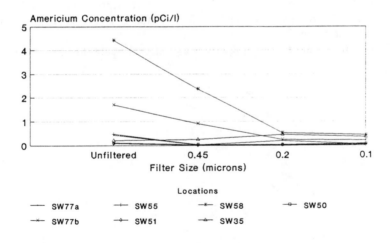

Figure 4. Americium-241 Concentration (pCi/l).

water analysis, field parameters, and description of the seeps are included in Appendix 1. The results of the ^{238}Pu were all near detection, except for Surface Water station SW-77 which showed some detectable concentrations in each size fraction. The results of the ^{239,240}Pu showed the main size fractions in the greater than 0.45 μm fraction, but with detectable levels in all of the size fractions. Figures 2 and 3 show these results.

The ^{241}Am results showed significant concentrations below the 0.2 μm and 0.1 μm fractions. This suggests a greater potential for colloidal americium relationships. Figure 4 shows results of the americium analysis. The results of the filter analysis were inconclusive because the volume of water passing through the filters was not recorded.

Additional studies are planned to further determine size distribution and chemical phases of potential plutonium/americium colloidal material using tangential-flow filtration in the range of 1.0 to 0.005 μm. Subsequent use of specially fabricated track-etched membranes will also be evaluated for better size distribution data. Sequential extractions of colloids will provide information on the mineralogical and chemical phases and the extent of plutonium and americium associations with these phases. In addition, scanning electron microscopy investigations will provide chemical and mineralogical information.

REFERENCES

U.S. Department of Energy, Rocky Flats Office, April 1990, Final Phase II RFI/RI Work Plan (Alluvial), 903 Pad, Mound, and East Trenches Areas, Operable Unit 2.

U.S. Department of Energy, Rocky Flats Office, Unpublished information and data.

Appendix 1. Plutonium and Americium Data for OU2 Seep-Water Filtrates.

Sample ID	Sample Date	Unfiltered Sample	Counting Error	0.45 um Filtrate	Counting Error	0.2 um Filtrate	Counting Error	0.1 um Filtrate	Counting Error
Americium-241 in pCi/L									
SW053	06/07/90	0.250	*(+-.0149)	0.570	*(+-.0274)	0.510	*(+-.0303)	0.400	*(+-.0178)
SW077	06/11/90	0.431	(+-.199)	0.030	*(+-.0385)	0.204	(+-.104)	0.037	*(+-.0329)
SW055	06/12/90	0.470	*(+-.0978)	0.033	*(+-.0384)	0.036	*(+-.0333)	0.072	*(+-.0490)
SW051	06/14/90	0.093	(+-.0495)	0.033	*(+-.0339)	0.031	(+-.0168)	0.030	*(+-.0307)
SW058	06/15/90	4.420	(+-.6700)	2.370	(+-.4400)	0.526	(+-.172)	0.436	(+-.138)
SW050	06/18/90	0.107	(+-.054)	0.033	*(+-.0329)	0.029	*(+-.0333)	0.076	(+-.0400)
SW077	07/02/90	1.700	(+-.3300)	0.911	*(+-.1900)	0.240	*(+-.149)	0.210	*(+-.1130)
Plutonium-239/240 in pCi/L									
SW053	06/07/90	0.313	(+-.0700)	0.225	(+-.054)	0.0332	(+-.0088)	0.018	*(+-.0351)
SW077	06/11/90	1.42	(+-1.31)	0.179	(+-.170)	0.152	(+-.146)	0.142	(+-.135)
SW055	06/12/90	2.21	(+-.4700)	0.0549	(+-.142)	0.108	(+-.027)	0.0935	(+-.0232)
SW051	06/14/90	0.029	*(+-.0265)	0.084	*(+-.0531)	0.14	(+-.132)	0.138	(+-.133)
SW058	06/15/90	22.3	(+-.3.900)	0.093	*(+-.0703)	0.405	(+-.093)	0.445	(+-.100)
SW050	06/18/90	0.154	(+-.146)	0.0488	(+-.0487)	0.023	*(+-.0235)	0.0409	(+-.0406)
SW077	07/02/90	3.32	(+/-.71)	0.18	*(+-.0736)	0.389	(+-.102)	0.0714	(+-.00186)
Plutonium-238 in pCi/L									
SW053	06/07/90	0.022	*(+-.0231)	0.034	*(+-.0348)	0.022	*(+-.0245)	0.046	(+-.00263)
SW077	06/11/90	0.026	*(+-.0302)	0.023	*(+-.0287)	0.010	*(+-.0266)	0.021	*(+-.0176)
SW055	06/12/90	0.02	*(+-.0261)	0.019	*(+-.0281)	0.034	*(+-.0302)	0.029	*(+-.0291)
SW051	06/14/90	0.023	*(+-.0219)	0.071	*(+-.0306)	0.018	*(+-.0199)	0.029	*(+-.0260)
SW058	06/15/90	0.053	(+-.0187)	0.074	*(+-.0617)	0.037	*(+-.0930)	0.0519	*(+-.0166)
SW050	06/18/90	0.018	*(+-.0150)	0.028	*(+-.0242)	0.019	*(+-.0218)	0.025	*(+-.0287)
SW077	07/02/90	3.32	(+-.7100)	0.180	*(+-.0736)	0.389	*(+-.1020)	0.0714	*(+-.0186)

Field Parameters						
Station ID	Date	Water Temp	Dissolved Oxygen (YSI)	Conductivity (mS/cm)	pH	Alkalinity
SW053	06/07/90	24	8.8	1.028	7.75	351
SW077	06/11/90	19	4.6	0.875	7.32	368
SW055	06/12/90	19	8.1	1.040	7.24	368
SW051	06/14/90	13	8.7	0.587	7.94	230
SW058	06/15/90	17	2.4	0.615	8.06	307
SW050	06/18/90	15	3.1	0.643	7.34	276
SW077	07/02/90	18	4.5	0.916	7.63	342

Legend: *Analytical Result is the Lower Limit of Detection for the Sample — No Data Available
mS/cm – Millisiemens Per Centimeter   YSI – YSI Dissolved Oxygen Meter

SW053 – Seep with Stagnant Water, Area Chocked with Vegetation, Sample Collected 30-ft South of Sample Marker
SW077 – Small Seep with Stagnant Water, Vegetation Around Site, 2 ft x 2 ft & 7-9 in. deep, Sample Taken in Front of Sample Marker
SW055 – Small Seep with Considerable Vegetation, 2 ft x 2 ft, Murky with algae Blloms, 6 in. deep
SW051 – Stram witn No Flow, Stagnant Puddle in front of Sample Sign, Considerable Algae, 5 in. Deep
SW058 – Small Seep Next to Roadside
SW050 – Swampy Seep, 6 in. X5 ft, Area Around Seep Full of Vegetation

CHAPTER 38

PCB Congeners as Tracers for Colloid Transport in the Subsurface: A Conceptual Approach

Remy J-C. Hennet and Charles B. Andrews

Polychlorinated biphenyls (PCBs) represent a group of 209 molecules containing 1 to 10 chlorine atoms distributed on the 10 available sites of a biphenyl molecular frame. Each different PCB molecule (also called PCB congener) has a set of specific chemical and physical properties, among them: aqueous solubility, octanol-water partition coefficient, and vapor pressure. These properties determine the mobility, fate, and behavior of PCBs in the environment. PCBs are found as contaminants in most natural environments, including soils and groundwater. In groundwater, PCBs are generally considered to be quite immobile because of their strong affinity to partition to the immobile organic content of soils. The presence of PCBs at a significant distance from a source area is usually explained by their migration (1) as part of a nonaqueous phase liquid (NAPL) or (2) during discrete events such as the overflowing of a disposal pit and subsequent surface runoff. However, PCBs sometimes occur away from a source area without any indication of NAPL migration or discrete run off events. Enhanced transport of PCBs sorbed on colloid-size particles, such as oil emulsions, humic or fulvic acids, clay, and quartz particles is being investigated.

The investigative concept is to use all PCB congeners as tracers of transport processes in the subsurface. Because of the specific properties of the different PCB molecules, it is possible to differentiate between dissolved transport and colloid transport based on the distribution of PCB congeners at the sampling location relative to the source location. Dissolved transport is characterized by the distributional predominance of the relatively more-soluble PCB congeners, whereas colloid transport will be

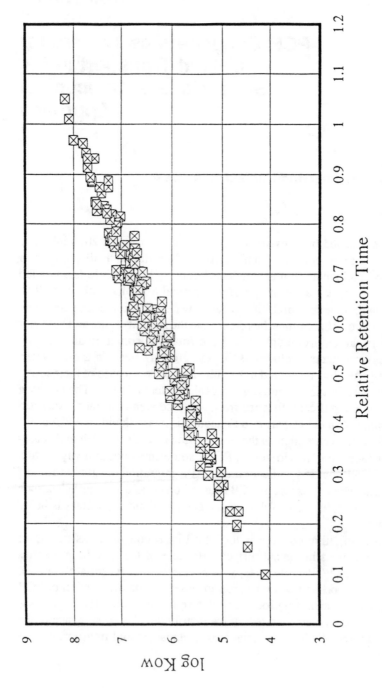

Figure 1.   Measured organic-water partition coefficients (logKow) versus relative retention times in a capillary column for different PCB cogeners (data from Hawker and Connell, 1988).

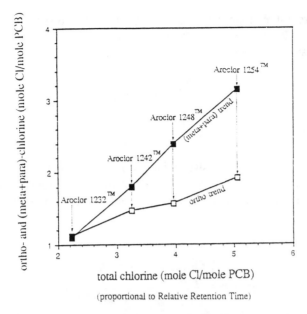

Figure 2. Measured chlorine substitution parameters for Aroclor products versus total chlorine.

characterized by the predominance of the more-hydrophobic and less-soluble PCB congeners. The study of the concept is greatly simplified by the fact that the specific properties of PCB congeners are all approximately linearly related to the relative retention time observed when PCB molecules are analyzed by capillary gas chromatography, as illustrated in Figure 1, for the octanol-water partition coefficient. Further simplification can be achieved by considering the average position of the chlorine substitutions (meta-, para-, and ortho-) for all the PCB molecules contained in a given sample, rather than the relative concentration of each one of the PCB congeners. This latest simplification step does not replace and carry over the entire set of information contained in the congener specific analysis but does permit the convenient characterization of a given sample by using only 3 rather than 209 parameters.

The chlorine substitution parameters (meta-, para-, and ortho-) are aligned when plotted vs total chlorine (the sum of meta-, para-, and ortho-chlorine substitutions) for the different PCB mixtures produced under the Monsanto trade name Aroclors⁻ in the United States (Aroclor 1232, 1242, 1248, and 1254) as illustrated in Figure 2 (the sum of meta- and para-chlorine substitutions is plotted as a single parameter). The reason for this alignment lies in the manufacturing and separation processes used to produce the different Aroclors⁻.

Two different retardation coefficients (R and R*) for dissolved transport of a single PCB congener i can be derived to characterize dissolved and colloid transport processes;

$$R_i = 1 + \frac{\rho_b}{n} \cdot foc \cdot e^{(A \log(K_{ow}) + B)}$$

where ρ_b stands for bulk density, n for porosity, foc for fraction organic carbon, and A and B are empirically determined coefficients (see, for example, Karickhoff et al. 1979). For colloid transport of a single PCB congener

$$R_i^* = 1 + \frac{\left[\dfrac{\rho_b}{n} \cdot foc \cdot e^{(A \log(K_{ow}) + B)} \right]}{1 + C' \cdot e^{(A' \log(K_{ow}) + B')}}$$

where C' represents the concentration of colloids, and A' and B' are empirically determined coefficients that depend on the nature of the colloids.

For a mixture of PCB molecules in a sample, the fingerprint of the sample, which is considered here has the three parameters meta-, para-, and ortho-, will be a function of the average of all the R's and R*'s of the PCB congeners present. For example, Figure 3a illustrates qualitatively the expected variations in the fingerprint parameters when dissolved transport dominates (case I) or when colloid transport dominates (cases II and III). Figure 3b shows qualitatively the changes in effective retardation (R or R*) for the three cases presented in Figure 3a; note that the effect of colloid transport is expected to be more important for the mobility of heavy PCB congeners (high relative retention times).

Additional work and information are required to fully calibrate the concept presented above for qualitative application at a given site. Among the site-specific parameters to be determined are the nature and concentration of colloids in the groundwater, as well as the seasonal variations in colloid concentration. However, for determining whether or not PCB contamination has reached a sampling point via dissolved or colloid transport, the concept can be applied simply by comparing the chlorine substitution parameters at the source and at the sampling location(s).

Results to date indicate that dissolved transport and colloid transport of PCBs in the subsurface can be differentiated based on the distribution of PCB congeners. Results also indicate that the concept can satisfactorily

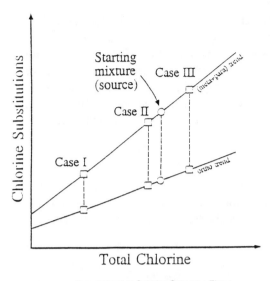

(proportional to Relative Retention Time)

Figure 3a. Chlorine substitution parameters and conceptual trends for dissolved and colloid transport processes. Case I: dissolved transport predominated; Cases II and III: colloid transport predominates. The reversed curvalure illustrated in Case III indicates that the light PCB cogeners can be preferentially partitioned to water (relatively to heavy PCB cogeners) during colloid transport.

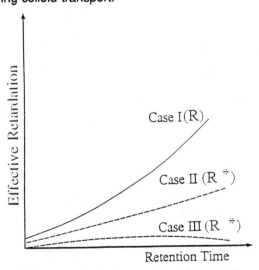

Figure 3b. Effective retardation versus retention time in a capillary column for PCB cogeners (see text and Figure 3a for explanation).

explain the presence of PCBs in a spring located more than 250 m from a source area and that the concept can also be used to differentiate transport processes for vertical migration of PCBs in the unsaturated and saturated zones. As an attempt to introduce quantification in the concept, mathematical modelling is being conducted in parallel with a long-term pumping test in an area contaminated with PCBs. A colloid tracer test will be performed in the spring of 1991 in the same area.

REFERENCE

Karickhoff, S.W., Brown, D.S., and Scott, T.A., Sorption of hydrophobic pollutants on natural sediments, *Water Res* 13:241-48, 1979.

CHAPTER 39

CTC - Colloid Transport Code and Simulation

Rohit Jain and H.E. Nuttall

The migration of toxic colloids in groundwater is being recognized as a potentially important factor in the assessment of waste sites. To assess the site- specific impact of toxic colloids on the environment mathematical models are needed; however, current contaminant transport models are inadequate to describe the migration of particles in groundwater. Therefore, the purpose of this study is to develop a general computer code (CTC-Colloid Transport Code) to solve colloid transport problems based on the population balance theory (Nuttall 1989a and 1989b; Randolph and Larson 1988; Travis and Nuttall 1987) and the Method of Lines as the numerical technique (Brown and Hindmarsh 1986; Hindmarsh 1984; Hyman 1976).

The modeling of colloid transport in groundwater is a complex problem requiring the solution of sets of nonlinear partial differential equations. To numerically solve these complex problems, we developed the Colloid Transport Code package (Jain 1991). This package is a set of Fortran 7 subroutines designed to solve the equations for colloid transport in saturated or unsaturated, porous or fractured media using the method of lines. The equations consist of the population balance equations for each colloidal species, the species balance equations for all components, and the energy and momentum balance equations. Thus, we may solve for the transport of components in the solute phase and as colloidal particles. In the current research version, the code can solve general systems of time-dependent second-order differential-algebraic equations in up to four spatial dimensions under a wide range of boundary and initial conditions. CRC uses the Method of Lines approach in that the spatial derivatives are discretized on uniform orthogonal grids forming a system of ordinary differential algebraic equations. These are solved using an efficient and robust

ODE solver package. The discretization is done by the package using a variety of finite-difference methods.

The user must provide input to a main program. Input parameters to the CTC code allow the user to select various discretization methods, to choose various ODE solving methods etc., and to set up the grids and the initial conditions. Also, the user must provide three subroutines for setting up the equations and the boundary conditions. All these are written in Fortran 77 and are very compact and simple to use. The package can interface with the NCSA Imagetool and Spyglass Dicer softwares for dynamic two- and three-dimensional simulations of colloid transport through saturated and unsaturated fractures under a varied set of boundary conditions. In the two-dimensional cases, single fractures were modeled assuming fully developed two-dimensional cases, single fractures were modeled assuming fully developed two-dimensional flow and a step input of colloids (size distribution). The boundary conditions at the walls of the fracture were varied to model various capture and release mechanisms. Initially, the two bounding cases were considered: (1) total capture at the walls and (2) a no-flux condition at the walls. Next, diffusion through the walls and into the rock were added to the model. After this, an electrokinetic model for adsorption-desorption at the wall was introduced. Unsaturated flow was also studies for the same cases, and then the case of total capture at the air-water interface was modeled. The three-dimensional case was developed to investigate the behavior of a distribution of colloid sizes in the same fracture. The flow was again assumed to by fully developed in two-dimension while the third dimension represented the size axis for the particles. The dependence of lateral diffusivity of colloids on size in conjunction with the width of the fracture determined the migration of a particular sized particle. Results for two cases are presented.

2D CASE: FRACTURE FLOW WITH DIFFUSION INTO THE TUFF AND SORPTION

- A narrow rectangular fracture contains colloids of a particular size flowing in groundwater. Colloids may interact with the surrounding rock matrix.

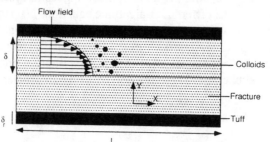

- Transport dominated by convection in the principal flow direction (x axis), while diffusion is dominant in the y axis.

- Neglect effects like agglomeration, birth, death, etc. Assume fully developed flow and neglect effects of surface-tension and flow of air (for the unsaturated case).

- Spherical particles (radius $r_p = 0.91$ m) in water at 20°C

$$D_y = \frac{2.13 \times 10^{-9}}{r_9 \ (in \ microns)} \ cm^2/s$$

- Adsorption/desorption, and diffusion into the tuff
  Physical parameters:
  $L = 90$ cm $\qquad\qquad u = c/c_o$
  u_s = surface concentration (per unit area)
  $\delta = 0.01$ cm $\quad \delta_r = 0.01$ cm $\quad (V_x)_{max} = 0.01157$ cm/s $\quad k_f = 0.1$
  $k_r = 10^{-4}$ $\qquad\qquad\qquad\qquad\quad D_x = 0.0011$ cm²/s
  $D_y = 2.44 \times 10^{-8}$cm²/s (colloid radius = 0.91 μm)

In the fracture,
$$\frac{\partial u}{\partial t} = - (V_x)_{max} \left(1 - \left(\frac{y}{\delta}\right)^2\right) \frac{\partial u}{\partial x} + D_x \frac{\partial^2 u}{\partial x^2} + D_y \frac{\partial^2 u}{\partial y^2}$$

while in the tuff,
$$\frac{\partial u}{\partial t} = 0.6 \ D_y \left(\frac{\partial^2 u}{\partial x^2} + \frac{\partial^2 u}{\partial y^2}\right)$$

and the electrokinetic adsorption/desorption model is
$$\frac{\partial u_s}{\partial t} = k_f \ u - k_r \ u_s.$$

Initial Conditions: $u = 0$.

Boundary Conditions:

at $x = 0$, $u = 1.0$ $\qquad\qquad$ at $x = L$, $\partial u/\partial x = 0$
at $y = 0$, $\partial u/\partial y = 0$ $\qquad\quad$ at $y = \delta + \delta_r$, $u = 0$.

Solution at T = 7,500 s:

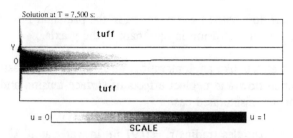

Solution at T = 7,500 s:

3D: CASE: FRACTURE FLOW WITH A SIZE DISTRIBUTION OF COLLOIDS

- A narrow rectangular fracture contains colloids of various sizes flowing in groundwater at a certain velocity. The colloids are instantly adsorbed into the rock matrix on coming in contact with it. In the principal flow direction, the transport is dominated by the convection term while in the other direction, diffusion is the dominant process.

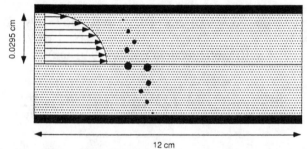

- Neglect effects like agglomeration, birth, death, etc. These terms can be easily incorporated into this example on a site-specific basis. The equation is,

$$\frac{\partial u}{\partial t} = -v_{av}\left(1 - \left(\frac{y}{\delta}\right)^2\right)\frac{\partial u}{\partial x} + D_x\frac{\partial^2 u}{\partial x^2} + D_y\frac{\partial^2 u}{\partial y^2}$$

The velocity profile above can be derived for a rectangular fracture. Here v_{av} (average velocity) = 2/3 v_{max} (maximum velocity). The parameters for the simulation are taken from representative data.

$$v_{max} = 0.0024 \text{ cm/s}$$

$$D_x = 0.0011 \text{ cm}^2/\text{s}$$

δ = half-width of the fracture = 0.0295 cm

Length of fracture = 12 cm.

- Assuming spherical particles and substituting values for water at 20°C into the Stokes-Einstein equation, we get

$$D_y = \frac{2.13 \ x \ 10^{-9}}{\text{radius (in microns)}} \ cm^2/s$$

If we were to include aggregation and other processes, we would use another axis as the size axis. This procedure will be followed here also since it makes data interpretation easier. Thus we have three independent axes. We will use a grid of 11 x 11 x 41 (NX x NY x NZ) where, as indicated in the previous example, we must use the axes as follows,

X ----> Size axis. We use logarithmic scaling from 1 nm to 1 μ.
Thus, $X = \ln(r/1\mu)/\ln(1nm/1\mu)$ where r is the radius of the particle.
Y ----> Width of the fracture, $0 \le Y \le 0.0259$ cm.
Z ----> Length of the fracture, $0 \le X \le 12$ cm.

Initial Conditions: u = 0
Boundary Conditions:

at X = 0, no boundary condition imposed (size axis).
at X = 1, no boundary condition imposed (size axis).
at Y = 0, $\partial u/\partial y = 0$ (symmetry).
at Y = δ, u = 0 (instant adsorption).
at Z = 0, u = 1.0 **(input PSD)**.
at Z = 12, u = 0 (at Z = infinity).

Solution at t = 4000 s.

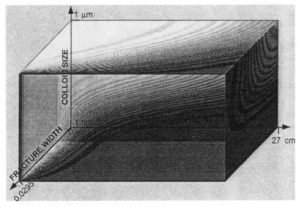

ACKNOWLEDGMENT

The research on which this publication is based was partially financed by the U.S. Department of Energy through the New Mexico Waste Management Education Consortium.

REFERENCES

Brown, P.N. and Hindmarsh, A.C. "Matrix-free Methods for Stiff Systems of ODEs," SIAM J. Numer. Anal. 23(3):610-38, 1986.

Hindmarsh, A.C., "ODE Solver for Time-Dependent PDE Software: Modules, Interfaces and Systems," B. Engquist and T. Smedsas (eds), 325-41, Elsevier Science Publishers B. V., North-Holland, IFIP, 1984.

Hyman, J.M. "The Method of Lines Solution of Partial Differential Equations," NYU Report COO-3077-139, 1976.

Jain, R., "CTCN: A User's Manual," unpublished MS thesis, Chemical and Nuclear Engineering Department, University of New Mexico, Albuquerque, NM, 1991.

Nuttall, H.E., "Interim Report on Particulate Transport," Milestone No. T424, TWS-EES-5/9-88-05, 1989a.

Nuttall, H.E., "Update Report on Colloid Transport: Milestone R528," NNWSI Milestone R528 (TWS-EES-5/10-87063), 1989b.

Randolph, A.D. and Larson, M.A., Theory of Particulate Processes, 2d ed., Academic Press, 1988.

Spyglass, Inc., 701 Devonshire Dr. C-17, Champaign, IL 61820.

NCSA Software Development, 152 Computing Applications Building, 605 East Springfield Avenue, Champaign, IL 61820.

Travis, B.J. and Nuttall, H.E., "Two-Dimensional Numerical Simulation of Geochemical Transport in Yucca Mountain," LA-10532-MS, UC-70, Los Alamos National Laboratory Report, NM, 1987.

CHAPTER 40

Characterization of Fine Colloid (<0.45 μm) Material Isolated from a Surface Soil from the Piedmont Region of North Carolina

Ruben M. Kretzschmar and Wayne P. Robarge

Colloid transport is known to occur in soils and subsoils. Early studies dealing with colloid transport were concerned with clay accumulation in the B horizons of soils and the formation of "clay skins" (Buol et al. 1980). More recent work has focused on a variety of areas involving colloid transport including: the influence of suspended colloids on soil hydraulic conductivity (Vinten et al. 1983a), formation of soil crusts (Helalia et al. 1988), movement of microbes (Smith et al. 1985; Wollum and Cassel 1978) and encapsulated agrochemicals (Lahav and Trapp 1980), and the transport of pesticides (Enfield et al. 1989; Vinten and Nye 1983b and 1985). The later work dealing with transport of pesticides demonstrates that transport of pollutants through soils via colloids is possible and may be enhanced for certain classes of pollutants such as hydrophobic compounds (Enfield et al. 1989). Under certain conditions it may be possible that colloid transport is even faster than water transport because of the exclusion of the colloid particles from the small pores (Enfield and Bengtsson 1989).

This poster presents preliminary results using a fine colloidal material (<0.45 μm equivalent diameter) isolated from a field-moist soil sample from a surface soil from the Piedmont region of North Carolina as a possible model for characterizing the behavior of colloidal particles in subsurface systems.

The colloidal material was isolated from an Ap horizon (plow layer) of an agricultural soil (clayey, kaolinitic, thermic, Typic Hapludult) from the Piedmont region of North Carolina.

Table 1:   Effect of dispersion treatment on release of clay (<2 μm) from an air-dry or field-moist soil.

Treatment	Air-Dry	Field-Moist
	- - - - - - - - - % Clay - - - - - - - - - -	
Distilled water	14.5	24.1
Hexameta-phosphate	19.2	24.8
Ultrasonic probe	21.5	25.9
Ultrasonic probe plus Hexameta-phosphate	24.7	26.7

Four dispersion treatments (distilled water, hexametaphosphate, ultrasonic probe, hexametaphosphate plus ultrasonic probe) were tested on air-dried and field-moist samples. Percent clay (<2 μm) as determined by the pipet method (Day 1965) was then compared (Table 1). Field-moist soil samples were very easily dispersed by shaking with distilled water. Since this method could be expected to result in the least chemical alteration of the particles, it was chosen for isolating the clay fraction <0.45 μm. A p - proximately 20 g of field-moist soil was shaken with 100 mL of distilled water for 24 h on a reciprocal shaker. The resulting suspension was diluted to 1 L, and the <0.45 μm (equivalent diameter) clay fraction was isolated by centrifugation (18.3 min at 2000 rpm, 100 mL centrifuge tubes, International No. II Centrifuge). The final concentration of the stock suspension was 800 mg clay per liter.

Critical coagulation concentrations (CCC) of the <0.45 μm clay fraction were determined using a procedure similar to the one described by Hesterberg and Page (1990). Suspensions of clay (40, 120, and 400 mg/L) were mixed with neutral salt solutions (0-5 mM CaCl$_2$, 0-250 mM KCl) and allowed to settle for fixed periods of time (3, 24, and 168 h). Clay particles remaining in suspension were determined by carefully removing 3 mL with a pipet from a depth of 3 cm and measuring absorbance (375 nm) on a spectrophotometer. The CCC was defined as the midpoint of the electrolyte concentration interval yielding the maximum changes in absorbance with increase in salt concentration (Hesterberg and Page 1990).

The percentage of clay remaining in suspension and the resulting CCC values were a function of initial clay concentration, time, and neutral salt concentration. Flocculation over relatively short time periods (3 h) required concentrated clay suspensions or high salt concentrations (Figure 1 and 2). The effect of initial clay concentration on percent clay remaining in suspension diminished with settling time. The CCC values in the pres-

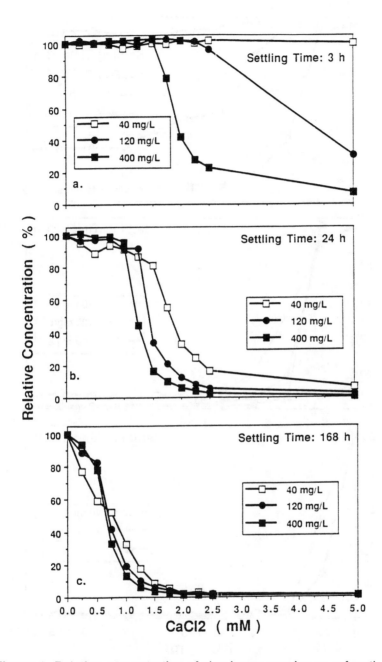

Figure 1. Relative concentration of clay in suspension as a function of time, neutral sale (CaCl$_2$) concentration, and initial clay concentration.

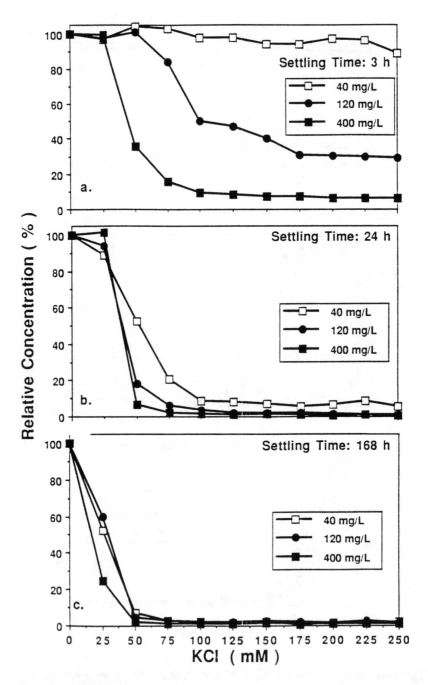

Figure 2. Relative concentration of clay in suspension as a function of time, neutral salt (KCl) concentration, and initial clay concentration.

ence of Ca^{2+} ranged from 0.6 to 1.8 mM. In the presence of K^+ CCC values ranged from 13 to 87 mM. The CCC values for either neutral saltcation or initial clay concentration continued to decrease with time (Figures 3 and 4).

The ability of the clay suspension to flow through a porous medium was tested using Ottawa fine sand (11.4% 250 to 500 μm; 4.6% 106 to 250 μm; 4.6% 53 to 106 μm) packed into a plastic column (diam 1.6 cm; length 20.2 cm; porosity 42%; one pore volume 16.5 mL). Clay suspensions (40 mg/L) were prepared by dilution of the stock suspension with distilled water. The flow rate (saturated flow) was maintained constant at 2.1 mL/min. Breakthrough curves of the clay suspension were determined by monitoring the total concentration of Fe and Al in the effluent with ICP Emission Spectroscopy.

The shape of the breakthrough curves suggested little restriction to the clay movement through the column (Figure 5). The degree of hydrodynamic dispersion was very small, probably because of the relatively coarse textured medium (fine sand) and the high flow rate (2.1 mL/min). However, continuous input of the clay suspension resulted in only 90% recovery. A visual layer of clay accumulation at the top of the sand column confirmed the loss of clay observed with Fe and Al analysis of the effluent. This layer of clay accumulation was probably due to coarser clay particles which could not penetrate the column.

The separated clay fraction (<0.45 μm) was predominately kaolinite with smaller amounts of hydroxy interlayered vermiculite. It also contained approximately 8% iron oxides, 5% gibbsite, and 2.5% total carbon (based on XRD, TGA, and TOC analysis). Acid extractable P (hot 6 N HCl) was 1.6 μg/mg clay.

The copper adsorption capacity of the fine clay was determined by equilibrating (24 h) aliquots of clay suspension (400 mg/L) with varying concentrations of $CuSO_4$ solutions, in the presence of 2 mM $CaCl_2$. The pH was adjusted by additions of HCl or NaOH. Copper adsorption was pH dependent and increased from no adsorption at pH 4.1 to ~4.7 mg Cu per gram of fine clay at pH 5.7.

The metal adsorption capacity of the fine clay fraction sets an upper limit on the amount of metal ions that can be transported by these colloids. For example, the maximum carrying capacity of a 40 mg Cu per liter clay suspension would be only 0.18 mg Cu per liter, assuming an adsorption maximum of 4.7 mg Cu per gram of clay at pH 5.7. With decreasing pH the Cu adsorption maximum decreases sharply, and simultaneously the Cu solubility increases. This suggests that the role of colloidal transport of heavy metals in soils may only be significant in very narrow pH ranges.

The ability of the fine clay particles to serve as a carrier for heavy metal transport was tested using a series of Zn^{2+} saturated cation exchange resins. Preliminary results indicate that these colloids do enhance Zn^{2+}

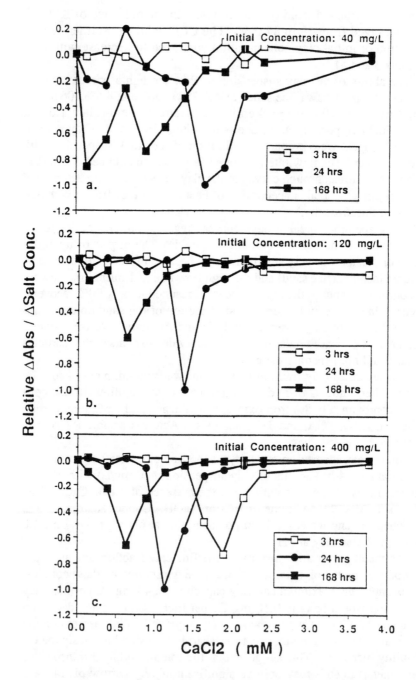

Figure 3. Relative Δ absorbance with increase in salt (CaCl$_2$) concentration as a function of time and initial clay concentration. The lowest ΔAbs/ΔSalt Conc. defines the CCC.

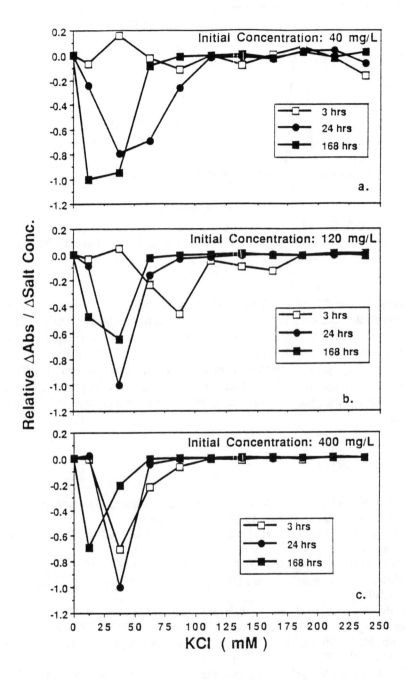

Figure 4. Relative Δ absorbance with increase in salt (KCl) concentration as a function of time and initial clay concentration. The lowest ΔAbs/ΔSalt Conc. defines the CCC.

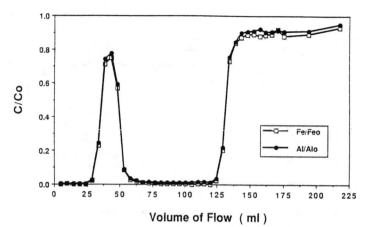

Figure 5. Breakthrough curves of a fine clay suspension (<0.45 µm; 40 mg/L) through a saturated sand column (distilled water; 2.1 ml/min).

removal from the resins. However, the total amount mobilized was relatively small compared to other reactions (e.g. cation exchange and hydrolysis) that were also occurring in the system. Work is proceeding on the further chemical and physical characterization of the fine clay fraction and on designing better systems to determine the extent of contribution of the colloidal material to pollutant transport. This work will be part of a larger study on the transport of solutes through saturated and unsaturated soil and subsoils.

REFERENCES

Buol, S.W., Hole, F.D., and McCracken, R.J., *Soil Genesis and Classification,* 2d Ed, Iowa State University Press, Ames, IA, 1980.

Day, P.R. and Black, C.A., et al. (eds.), Methods of soil analysis, Part I, *Agronomy* 9:545-67, 1965.

Enfield, C.G. and Bengtsson, G., *Ground Water* 26:64-70, 1988.

Enfield, C.G., Bengtsson, G., and Lindqvist, R., *Environ Sci Technol* 23:1278-86, 1989.

Helalia, A.M., Letey, J., and Graham, R.C., *Soil Sci Soc Am J* 52:251-55, 1988.

Hesterberg, D. and Page, A.L., *Soil Sci Soc Am J* 54:729-35, 1990.

Smith, M.S., Thomas, G.W., White, R.E., and Ritonga, D., *J Environ Qual* 14:89-91, 1985.

Vinten, A.J.A., Mingelgrin, U., and Yaron, B., *Soil Sci Soc Am J* 47:408-12, 1983a.

Vinten, A.J.A. and Nye, P.H., *J Soil Sci* 36:531-41, 1985.

Wollum, A.G. and Cassel, D.K., *Soil Sci Soc Am J* 42:72-78, 1978.
Lahav, N. and Trapp, D., *Soil Sci* 130:151-56, 1980.

CHAPTER 41

Iron Dynamics During Injection of Sandy Aquifer

Liyuan Liang, John F. McCarthy,
Thomas M. Williams, and Lou Jolley

The dynamics of dissolved, colloidal and solid phase iron were examined during a forced-gradient field experiment in Georgetown, South Carolina, in the summer of 1990. The experiment involved injection of 80,000 L of oxygenated water containing high levels of natural organic matter (66 mg/L NOM) into a sandy aquifer (McCarthy et al. 1992). Iron dynamics were followed in three saturated horizons at sampling wells located 1.5 m (A wells) and 3 m (B wells) from the injection well (see Figure 1, McCarthy et al. 1992). The initial oxidation/reduction potential of the aquifer favored Fe(II) in the iron-rich groundwater.

As oxygen-rich water is introduced into the groundwater, the redox potential was expected to increase. The impact of the changing redox potential on iron dynamics was hypothesized as following: (1) Fe(II) may decrease as an expense of Fe(III) production by oxygenation, but otherwise be mobile and conservative; (2) Fe(III) is mostly in the ferric oxide/hydroxide colloidal fraction (i.e., Fe(III) may be found predominantly in sizes greater than about 1 nm (>3000 molecular weight, or 3K), and the transport of colloidal iron oxide may be limited; (3) ferric oxide/hydroxide colloids may have positive or near-zero surface charge in the pH range of the groundwater(6.0 to 7.2); and (4) the turbidity of groundwater may increase as a result of the formation of iron colloids.

During the course of injection, dissolved oxygen (D.O.), pH, iron concentration, NOM, and turbidity were measured. Samples were also collected for micro-electrophoretic mobility measurement (Coulter model: Delsa 440), particle size analysis (Coulter model: N4MD, based on PCS), and scanning electron microscopic examination (ISI-40).

**CHANGES IN D.O. AND TURBIDITY
DURING NOM INJECTION**

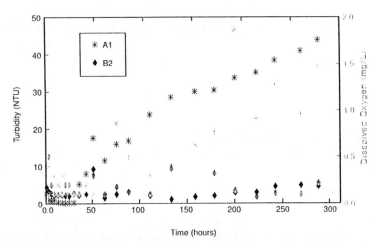

Time (hours)

Figure 1. Changes in dissolved oxygen and turbidity during injection for sampling well A1 and B2. Injection started on 7/10/90 and ended on 7/23/90.

Results from the **A1 well** - the sampling well located in the zone closest to the water table and in the 1.5 m sampling well cluster, showed that D.O. increased substantially from 0.1 to 1.5 mg/L and turbidity increased by a factor of 10 during the injection period (Figure 1). A comparison of ferric/ferrous concentration and size distribution at the beginning and the end of injection shows that Fe(III) was increased more than 20-fold. Fe(II) in the same period was reduced to a tenth of its initial value. At the end of the injection (Figure 2), 73% of Fe(III) was retained on a 0.1 µm filter. Fe(II), however was mainly in the dissolved form (passing through 3K and 100K cut filters). The organic matter was largely in <3K size range, with about 30% of organic carbon in the range of 3K to 100K size range.

The electrophoretic mobilities of groundwater particles in the A1 well at the beginning and the end of the NOM injection both showed negative surface potentials. The amount of groundwater particles increased substantially as indicated by counts per second (CPS) from 8.4×10^3 to 148×10^3 at the beginning and the end of the injection, respectively. The sizes of the groundwater particles fell into two ranges: one about 200 nm and the other >3 µm. This is in good agreement with the Scanning Electron Micrograph (SEM), which showed individual spherical particles of 200 nm and large aggregates of these particles. X-ray analysis on SEM samples indicated the presence of iron.

SIZE DISTRIBUTION OF Fe(II), Fe(III)
AND NOM AT END OF NOM INJECTION

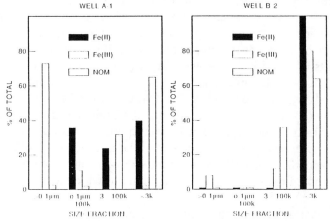

Figure 2. Size distribution of Fe(II), Fe(III) and dissolved organic carbon at the end of injection. Amicon hollow fiber filters were used for each size cut.

Because A1 well is close to the water table, moderately high D.O. content was achieved at the end of the injection (1.5 mg/L). In the presence of oxygen, the Fe(II) is rapidly oxygenated, and the rate is on the order of minutes, as demonstrated in laboratory study (Sung and Morgan 1980). The disappearance of Fe(II) and the production of Fe(III) during the injection may be accounted by the fast oxygenation process.

Because Fe(III) hydroxides/oxides are sparingly soluble at the groundwater pH, precipitation of Fe(III) will follow oxidation. The observation from SEM and PCS support the existence of colloidal groundwater particles, and these colloids are most likely iron colloids because most of Fe(III) is in the size range greater than 0.1 μm. Colloids of >0.1 μm will be very effective in scattering light and influencing turbidity.

At pH 6.0-7.2, most iron hydroxide/oxide particles will have a positive or neutral charge (James and Parks 1982). The consistent negative surface potential from electrophoretic mobility measurement suggests that iron colloids are associated with anions. Fourier transfer infrared spectroscopic analysis (FTIR) on a sample from A1 showed a strong spectrum for -COO⁻ (Dr. N. Marley, Argonne Laboratory, personal communication). Carboxylate bonding to iron oxide has been established by Peck et. al. (1966), hence the association of iron colloids with organic anions may result the negative surface potential of particles.

The lack of positively charged particles can also be interpreted as follows: because the aquifer material is composed of negatively charged sand material, any positively charged Fe particles will be preferably collected.

Consequently, the transport of iron colloids would not be possible. However, the increase in Fe(III) is twice as great as the decrease of Fe(II) in the sampling well at the end of the injection, suggesting that transport of iron colloids may be occurring. This apparent transport may be the result of the mobilization of the anion coated Fe colloids in the groundwater.

Results from the **B2 well**—the sampling well located in a slow-flowing zone, 3 m from the injection well, showed that both turbidity and D.O. remained constantly low during the NOM injection (Figure 1). A comparison of Fe(II) and Fe(III) distribution at the beginning and the end of the NOM injection shows that Fe(III) concentration was doubled, whereas Fe(II) increased slightly. At the end of the injection (Figure 2), 80% of Fe(III) and 100% of Fe(II) were in the <3K size cut. The organic matter is distributed mainly in the <100K range, in which 62% is in <3K cut range.

The electrophoretic mobilities of groundwater particles showed only negative surface potentials. An increase in the CPS is observed over injection period. On SEM, large aggregated particles can be identified, and X-ray analysis of the 200 nm particles indicated iron.

The B2 well lies in a zone of slow groundwater movement. Dissolved oxygen injected with the NOM water may be used up in the oxygenation process before reaching B2. Alternatively, D.O. may be taken up from microbial activities. These may cause the low D.O. during the injection. Under low D.O. and high NOM conditions, Fe(II) oxygenation is limited. Fe(II)-Fe(III) may act as a catalyst for oxidation of organic material (Stumm and Morgan 1981), as explained by the following reactions:

Fe(II) + organic --> Fe(II)-organic complex

Fe(II)-organic complex + O2 -->Fe(III)-organic complex

Fe(III)-organic complex---> Fe(II) + oxidized organic

The observation of high ferrous concentration throughout the injection may be a result of a slow oxygenation of Fe(II) in comparison to the Fe(III) reduction by organic matter. Alternatively, bacteria may use Fe(III) as an oxidizing agent in the breakdown of organic carbon, producing Fe(II).

In the B2 well, levels of both Fe(II) and Fe(III) increased during the injection, but most of the iron was in the <3K size cut. This may be accounted for by the high NOM enhancing iron colloid dissolution and the formation of complexes of low-molecular-weight NOM compounds with the Fe(III), rather than the Fe(III) precipitates observed in the A1 well. The very low concentration of Fe(III) of greater than 0.1-μm size results in little intensity of light scattering and is consistent with the observed low turbidity. The formation of large aggregates as identified in SEM may be a result of the bridging between organic matter and iron colloids (>0.1 μm).

It is concluded that oxidation of Fe(II) to Fe(III), followed by precipitation of ferric hydroxide/oxide, accounts for the turbidity increase in the A1 well. The interaction of NOM and iron oxide particles alters the colloid surface electric charge/potential. NOM thereby stabilizes iron colloids and allows the transport of inorganic colloids in porous media.

In B2 well, oxygenation of Fe(II) is slow because of low D.O., thus maintaining a steady-state Fe(II) concentration. Increased concentration of both Fe(II) and Fe(III) are postulated as being the result of the NOM-enhancing iron colloid dissolution and the formation of Fe-organic complexes, rather than the precipitation of Fe(III) as was observed in the A1 well.

Currently, analysis is in progress on cation/anion speciation and organic chemistry. New information will be of great help in interpreting the observations.

REFERENCES

James, R.O. and Parks, G.A., Characterization of aqueous colloids by their electrical double-layer and intrinsic surface chemical properties, *Surface Colloid Sci* 12, Matijivic Ed., Plenum, New York, 1982.

McCarthy, J.F., Liang, L., Jardine, P.M., and Williams, T.M., Mobility of natural organic matter injected into a sandy aquifer, Summary Report Concepts in *Manipulation of Groundwater Colloids for Environmental Restoration*, Lewis Publishers, Inc., Chelsea, MI, 1992.

Peck, A.S., Raby, L.H., and Wadsworth, M.E., An infrared study of the flotation of hematite with oleic acid and sodium oleate, *Trans AIME* 235:301, 1966.

Stumm, W. and Morgan, J.J., in Chapter 7, *Aquatic Chemistry*, Wiley and Son, New York, 1981.

Sung, W. and Morgan, J.J., Kinetics and product of ferrous iron oxygenation in aqueous systems, *Environ Sci Technol* 14:561, 1980.

CHAPTER 42

Mechanism of Detergency for Polar Oils: Is it Relevant to Removal of Organic Contaminants from Soils?

Jong-Choo Lim and Clarence A. Miller

Removal of oily soils from synthetic fabrics occurs by a solubilization-emulsification mechanism. For example, our previous work using videomicroscopy has shown that hydrocarbons are solubilized into microemulsion phases that are not present initially but develop during the washing process near the initial surface of contact between surfactant solution and oil. The low interfacial tensions associated with microemulsions facilitate their subsequent emulsification into the washing bath. In recent work, we have shown that pure polar oils such as long-chain alcohols and mixtures of such compounds with hydrocarbons are solubilized into liquid crystalline phases that, similarly, form near the initial surface of contact. A theory has been developed to describe the diffusion process leading to liquid crystal formation. One may ask whether oily materials on the surfaces of soil particles may be removed by a similar mechanism.

Detergency is among the oldest applications of surfactants and remains the largest in terms of surfactant usage. Although much is known about detergency, continuing research to obtain a better understanding of its mechanisms is necessary because of ever increasing demands on cleaning products. For instance, the current widespread use of synthetic fabrics and the associated use of lower washing temperatures have been the main cause of a substantial increase in recent years in the amount of nonionic surfactants used in household laundry products. The new products, which are mixtures of nonionic and anionic surfactants, are more effective than anionics alone in removing oily soils from synthetic fabrics, apparently because they operate by a mechanism involving solubilization and emulsification. This behavior contrasts sharply with the removal of oily soils from cotton during high-temperature washing with anionic surfactants, which is widely believed to be by the rollback mechanism. During rollback, the surfactant solution makes the fabric more water wet and displaces the liquid soil from the solid. Rollback seems not to be the dominant mechanism for synthetics, however, probably because adhesion between solid and soil is greater than with the more polar cotton.

Central to understanding the mechanism of detergency is knowledge of the dynamic behavior that occurs when an aqueous surfactant solution contacts an oily soil. We have employed videomicroscopy to make direct observations of this behavior in model systems for which key features of equilibrium phase behavior have been determined (Benton et al. 1986, Raney et al. 1987; Raney and Miller 1987; Mori et al. 1989, 1990; Shinoda and Friberg 1975). Nonionic surfactants or mixtures of nonionic and anionic surfactants were used in these experiments. The soils were liquid hydrocarbons, triglycerides, long-chain alcohols, and mixtures of these materials. Results of the videomicroscopy experiments were compared with results of fabric washing tests conducted at Shell Development Company using the same model systems. The main conclusion regarding detergency derived from these studies is that good soil removal occurs when intermediate phases that were not present initially and that solubilize considerable soil develop on contact of surfactant solution with soil. These intermediate phases are typically microemulsions or liquid crystals.

A new contacting technique was developed to observe the dynamic behavior of small, individual drops of oily soils immersed in a large quantity of surfactant solution. With this technique, which differs from that of our previous contacting studies in which comparable volumes of oil and surfactant solution were employed (Benton et al. 1986, Raney et al. 1987; Mori et al. 1989) the latter stages of solubilization, a significant part of the detergency process in these systems, can be observed. Moreover, partitioning of various components between phases, an important phenomenon in systems containing mixed soils and/or surfactants and one dependent on the relative amounts of surfactant solution and soil contacted, can be made close to that expected in practical systems.

Samples of the surfactant solutions to be studied were introduced into rectangular optical glass capillaries (Vitrodynamics, Inc.) by capillary action. The capillaries were 50 mm long and 4 mm wide and had an optical path length of 400 µm. For cleaning, they were soaked in chromic acid solution for at least 24 h, rinsed several times with distilled water, and dried before use. After filling, they were sealed at one end and attached to standard microscope slides with UV-light-sensitive polymer adhesive, which was cured with an ultraviolet fiber optic gun. A very thin hypodermic needle (0.21 mm OD, Hamilton Company) was used to inject individual oil drops into the surfactant solution (see Figure 1). This procedure

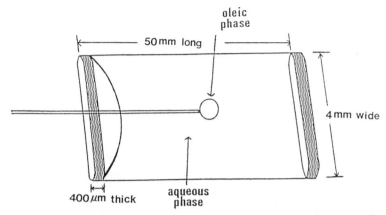

Figure 1. Schematic illustration of oil drops contacting experiment where a small oily drop is injected in aqueous surfactant solution.

gave good control of drop size. Drops about 100 µm in diameter were injected.

Temperature control for the samples was obtained with a Mettler FP-52 controller and FP-5 microscope hot stage. The thermal stage was modified and placed on top of a new X-Y positioner (Nikon, Inc.) so that drops could be observed during and immediately after injection while the sample was maintained at constant temperature.

A quasi-steady-state approach was used to analyze the experiment. Details are given elsewhere (Lim and Miller 1991). Basically, the analysis indicates that drop composition advances with time along the coexistence curve, as shown in Figure 2, until it reaches point, which is one vertex of a three-phase triangle. At this time the lamellar liquid crystalline phase, which is the third phase involved, begins to form. According to the analysis, the time t until liquid crystal formation is given by

$$t = K_s \left(\frac{R_0^2}{D_2 w_{2\infty}} \right)$$

where D_2 is a diffusion coefficient of surfactant, $w_{2\infty}$ is initial surfactant concentration, and R_0 is the radius of the initial oil drop.

Equation (1) predicts that the time required to initiate liquid crystal formation is proportional to the square of the initial drop size and inversely proportional to the initial surfactant concentration and the surfactant diffusion coefficient. K_s is a constant that includes the effects of the shape of

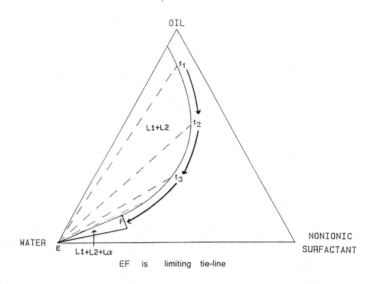

Figure 2. Schematic pseudoternary phase diagram of nonionic surfactant-water-mixture of n-16$_{16}$ and oleyl alcohol showing change in oil drop composition with time where EF is the limiting tie-line.

the $(L_1 + L_2)$ coexistence curve and the location of the limiting tie-line EF (see Figure 2).

It is noteworthy that the preceding analysis does not require extensive information on phase behavior (e.g., on the single phase and multiphase regions involving the liquid crystalline phase). All that is needed is the coexistence curve between the aqueous surfactant solution L_1 and the oil phase L_2 for the temperature and oil composition of interest. This feature is especially significant for multicomponent mixed soil systems, where the effort required to determine a complete phase diagram is enormous.

Figure 3 is a video frame showing an intermediate liquid crystalline phase formation for the system containing 0.05 wt% $C_{12}E_8$ nonionic surfactant solution with 85/15 n-C_{16}/oleyl alcohol at 50°C, where the initial drop has a diameter of about 50 μm. During the next minute, myelinic figures grow rapidly into the surrounding aqueous phase and the oil is incorporated very rapidly into the liquid crystal during the process.

Raney and Benson (1990) reported soil removal results in mixed non-polar-polar soil systems. Based on our videomicroscopy observations in several of the same systems, it is considered that the good detergency is caused by the same mechanism as that discussed in the preceding paragraph (i.e., the formation of a liquid crystalline intermediate phase as

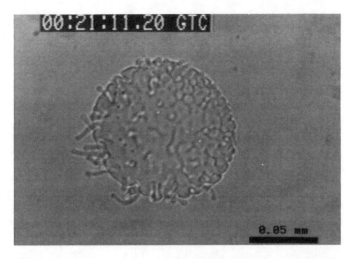

Figure 3. Video frames showing formation of intermediate lamellar liquid crystalline phase following contact of 0.05 wt % $C_{12}E_8$ surfactant solution with drop of 85/15 n-C_{16}/oleyl alcohol at 50°C.

myelinic figures). That is, excellent soil removal is found for the systems where a liquid crystalline phase that solubilizes substantial portions of the oil forms within a short time after surfactant solution and soil are brought into contact.

This approach shows that surfactant solutions can remove polar compounds and their mixtures from fabrics by solubilization into liquid crystal. Perhaps this process could be used to remove some oily materials from contaminated soils or aquifers.

REFERENCES

Benton, W.J., Raney, K.H., and Miller, C.A., *J Colloid Interface Sci* 110:363, 1986.

Lim, J.C. and Miller, C.A., *Langmuir* 7:2021, 1991.

Mori, F., Lim, J.C., and Miller, C.A., *J Amer Oil Chem Soc* 67:22, 1990.

Mori, F., Lim, J.C., Raney, O.G., Elsik, C.M., and Miller, C.A., *Colloids Surf* 40:323, 1989.

Raney, K.H. and Benson, H.L., *J Amer Oil Chem Soc* 67:22, 1990.

Raney, K.H. and Miller, C.A., *J Colloid Interface Sci* 119:539, 1987.

Raney, K.H., Benton, W.J., and Miller, C.A., *J Colloid Interface Sci* 117:282, 1987.

Shinoda, K. and Friberg, S., *Adv Colloid Interface Sci* 4:281, 1975.

CHAPTER 43

Retardation of Dissolved Soil Organic Matter and its Influence on Mobility of Polycyclic Aromatic Hydrocarbons (PAH)

*Leonard W. Lion, Brian R. Magee, and
Ann T. Lemley*

The retardation factor of phenanthrene in a sand column was reduced by an average factor of 1.8 in the presence of dissolved organic matter (DOM) derived from soil. Enhanced transport of phenanthrene was observed in spite of the fact that the transport of DOM was retarded relative to ^3H_2O. A simple three-phase model was developed for calculation of pollutant retardation in the presence of a carrier that sorbs to the porous medium. Pollutant retardation factors in the presence of a carrier may be reasonably predicted based on independent measurement of distribution coefficients for (1) the pollutant with the porous medium, (2) DOM with the porous medium, and (3) the pollutant with DOM.

Because the binding of organic pollutants and carrier retardation are both likely to be enhanced by increasing DOM hydrophobicity there appear to be practical limits on the extent to which DOM can enhance PAH transport.

It is relatively well established that if dissolved macromolecules or colloidal particles are dispersed in the groundwater, they may sorb hydrophobic pollutants and alter their transport. Enhanced or "facilitated" transport will result if the colloid-bound pollutant is more mobile than the dissolved pollutant in the absence of a colloidal carrier. For colloids or dissolved macromolecules to increase pollutant mobility, two conditions must be met: (1) the "carrier" (i.e., colloid or dissolved macromolecule) must be able to bind the pollutant to an appreciable extent and (2) the carrier must have a greater mobility than the pollutant which it binds. Given these constraints, it is interesting that most studies of potential carriers have focused on their pollutant binding properties and neglected to consider

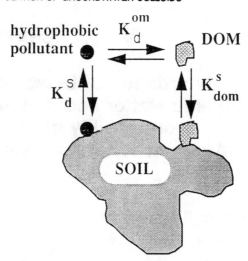

Figure 1. Schematic of the possible interactions between a hydrophobic pollutant, a carrier (dissolved soil organic matter) and the stationary phase (soil). (Reprinted with permission from Magee et al., 1991. Copyright American Chemical Society.)

their mobility. In models of the effect of colloids on pollutant mobility it is often assumed that the carrier moves with a velocity equal to or greater than that of the pore water, that is, that carrier is not retarded and/or is excluded from pores (West 1984; Hutchins et al. 1985; and Enfield et al. 1989).

In the case where the carrier is an organic macromolecule, it is plausible to assume that the carrier will reversibly bind to the stationary porous media and be retarded relative to the pore water velocity. If such a sorption reaction occurs, the ability of the carrier to enhance pollutant transport will be diminished. Both the binding of a pollutant to a carrier and the interaction of the carrier with the porous media can be expected to increase with increasing carrier hydrophobicity. A tradeoff is, therefore, likely to exist between carrier properties that enhance pollutant solubility (high carrier hydrophobicity) and those that enhance carrier mobility (low carrier hydrophobicity). The objective of the research described here was to evaluate the extent to which transport of a hydrophobic pollutant may be enhanced by carriers that are subject to retardation.

The interactions between a carrier, pollutant, and the stationary phase are illustrated in Figure 1. A model that describes the retardation of a pollutant in the presence of a carrier that partitions on to the stationary phase has been developed by Magee et al. 1991. The resulting retardation factor $R*$ is a function of the pollutant partition coefficient with the stationary phase,

K_d^s, the pollutant partition coefficient with the carrier, K_d^{om}, as well as the partition coefficient of the carrier with the stationary phase, K_{dom}^s, and is given by

$$R^* = \frac{(1 + K_d^{om} DOM + K_d^s \rho_b/n)}{1 + \dfrac{K_d^{om} DOM}{1 + (K_{dom}^s \rho_b/n)}}$$

where ρ_b = the bulk density of the porous medium, n = the effective porosity (water content), and DOM is the aqueous concentration of mobile organic matter.

If the interactions illustrated in Figure 1 are sufficient to describe behavior of the hydrophobic pollutant in the presence of a carrier, then the individual distribution coefficients K_d^{om}, K_d^s, and K_{dom}^s determined in independent experiments should predict the behavior of pollutant-carrier mixtures. This hypothesis was tested in this research using data obtained from batch and column experiments.

Because the hydrophobicity of the carrier will control both K_d^{om} and K_d^{om}, it is reasonable to expect that they may covary. For purposes of illustration, consider a case in which DOM = 100 mg/L, K_d^s = 12.9 mL/g (the value determined for the partition coefficient for phenanthrene onto the sand used in the experiments described below), ρ_b/n = 4.38 g/mL (the value for the experimental sand), and the ratio of K_d^{om} to K_{dom}^s is constant. For this hypothetical case, R* may be determined from Eq. (1) as a function of K_d^{om}. The results are shown in Figure (2).

Figure 2 shows that a carrier sorption coefficient exists that would provide a minimum R* (the reduction in R* is less apparent at lower DOM concentrations and more apparent at higher ratios of K_d^{om} to K_{dom}^s). The minimum in the retardation factor would represent the optimal condition for the enhancement of pollutant transport. Extremely hydrophobic carriers (i.e, those with large K_d^{om}) will act to increase the retardation of the pollutant above that observed when the carrier is absent.

[14]C-labeled phenanthrene was used in all experiments and assayed by liquid scintillation counting. A sand obtained from a quarry in Newfield, New York, was used as the stationary phase in column experiments and as the sorbent in batch experiments. Solutions of DOM were prepared by extraction from a clayey silty loam (A-horizon; Rhinebeck, New York) with 0.005 M $CaSO_4$ plus 1.0 g/L NaN_3, (@1:5 soil:solution ratio). The extract was centrifuged @ 5400xg (30 min) and then filtered through a Whatman 2V filter. A typical DOM extract contained 80 mg/L DOC.

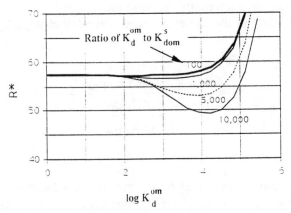

Figure 2. Variation in Pollutant Retardation Factor as a Function of the Carrier Hydrophobicity.

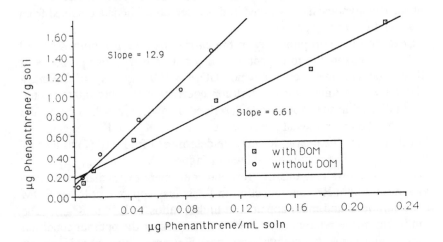

Figure 3. Phenanthrene sorption isotherms onto sand in the absence and presence of dissolved soil organic matter. (Reprinted with permission from Magee et al., 1991. Copyright American Chemical Society.)

Distribution coefficients were determined by batch and column methods for phenanthrene and DOM with the sand. Phenanthrene binding to DOM was measured by fluorescence quenching using a procedure similar to that of Gauthier et al. (1986).

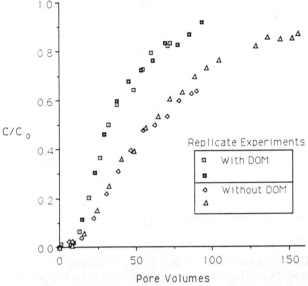

Figure 4. Phenanthrene breakthrough curves in the absence and presence of dissolved soil organic matter. (Reprinted with permission from Magee et al., 1991. Copyright American Chemical Society.)

Sorption of phenanthrene was reduced in batch experiments by the presence of DOM (see Figure 3) and phenanthrene mobility in a sand column was increased by the presence of DOM (see Figure 4). DOM sorption by the sand obeyed a linear isotherm, and it was retarded relative to an inert tracer (^3H_2O) in transport through the sand (see Figure 5).

Phenanthrene binding by DOM also obeyed a linear isotherm (see Figure 6). Note that Gauthier et al. (1986) have shown that the binding constant K_d^{om} can be related to the phenanthrene fluorescence intensities in the absence and presence of DOM (F_0 and F, respectively) by:

$$\frac{F}{F_0} = K_d^{om} [DOM] + 1$$

This relationship is used to obtain K_d^{om} in Figure 6.

If the simple model shown in Figure 1 is reasonable, then pollutant retardation in a system containing a carrier (also subject to retardation) can be predicted using Eq. 1. Distribution coefficients for phenanthrene with the sand, DOM with the sand, and phenanthrene with DOM determined from the separate batch and column experiments were used to predict aggregate behavior of phenanthrene in the presence of DOM.

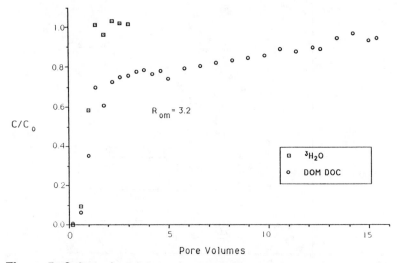

Figure 5. Column breakthrough curves for dissolved soil organic matter and $^3H_2$. (Reprinted with permission from Magee et al., 1991. Copyright American Chemical Society.)

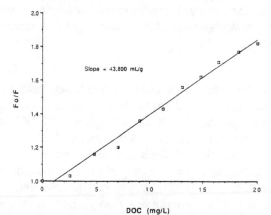

Figure 6 Fluorescence quenching of phenanthrene by dissolved soil organic matter. (After Magee et al., 1989.)

Overall, independent column studies with DOM and phenanthrene produced the best prediction of aggregate (phenanthrene plus DOM) behavior in columns. The distribution coefficients determined by the separate column studies with phenanthrene and DOM resulted in a predicted retardation factor (R*) of 37 for phenanthrene in the presence of DOM vs an observed value of 44 (within 16%).

The experimental observations confirm that a carrier that is retarded in porous media can still enhance the transport of a hydrophobic pollutant. For a carrier to be effective in enhancing pollutant transport in porous media it must meet two criteria: (1) it must effectively bind the pollutant in solution [higher values of K_d^{om} in Eq. (1)] and (2) it must be mobile [lower values of K_{dom}^s in Eq. (1)]. It seems reasonable to expect that these two factors will not be independent because they both relate to the hydrophobic nature of the carrier. Increased K_d^{om} will likely be reflected in increased K_{dom}^s. Several investigations have shown that humic materials can bind hydrophobic pollutants; however, the mobility of these materials in porous media must also be considered.

REFERENCES

Enfield, C.G., Bengtsson, G., and Lindqvist, R., Influence of macromolecules on chemical transport, *Environ Sci Technol* 23:(10)1278-86, 1989.

Gauthier, T.D., Shane, E.C., Guerin, W.F., Seitz, W.R., and Grant, C.L., Fluorescence quenching method for determining equilibrium constants for polycyclic aromatic hydrocarbon binding to dissolved humic materials, *Environ Sci Technol* 20:1162-66, 1986.

Hutchins, S.R., Tomson, M.B., Bedient, P.B., and Ward, C.H., Fate of trace organics during land application of municipal wastewater, *CRC Crit Rev Env Control* 15:355-416, 1985.

Magee, B.R., Lemley, A.T., Lion, L.W., The effect of water soluble organic material on the transport of phenanthrene in soil, in *Petroleum Contaminated Soils,* Calabrese E.J. and Kostecki, P.T., (eds.), 2:157-76, 1989.

Magee, B.R., Lion, L.W., and Lemley, A.T., The transport of dissolved organic macromolecules and their effect on the transport of phenanthrene in porous media, *Environ Sci Technol* 25(2):323-31, 1991.

West, C.C., Dissolved organic carbon facilitated transport of neutral organic compounds in subsurface systems, Ph.D. Thesis, Rice University, 1984.

CHAPTER 44

Colloid Transport in Packed-Bed Columns with Repulsive Energy Barriers

Gary M. Litton and Terese M. Olson

The mobility of colloids and colloid-associated contaminants in subsurface environments is dependent on the deposition rates of these particles from solution to collector surfaces. When particle deposition is inhibited by a repulsive interaction energy barrier between the colloid and the collector, existing theories have been shown to poorly predict particle transport (Bowen and Epstein 1979; Elimelech and O'Melia 1990a, 1990b; Gregory and Wishart 1980; Tobiason 1989). A number of possible explanations for this phenomenon were recently proposed by Elimelech and O'Melia (1990a, 1990b). The work presented here suggests that the differences between theoretical and experimental behavior are partially attributable to the cleanliness of the surface collectors. Filtration experiments suggest that colloid deposition rates on well-cleaned quartz or glass beads exhibit behavior that is more consistent with theory; however, exact predictions of attachment efficiencies have not been obtained.

Classical Derjaguin-Landau and Verwey-Overbeek (DLVO) theory of colloid stability provides a surface chemistry model that is commonly combined with transport theory to predict colloid mobility in porous media. It considers attractive London-van der Waals forces and repulsive electrical double-layer interactions. Unfavorable filtration conditions exist when electrical double-layer forces are sufficient to create a repulsive energy barrier between the particles and the collector medium. This may occur when the particles and the collectors are either both positively or negatively charged.

Deposition rates were expressed in terms of attachment efficiencies, α, where

$$\alpha = \frac{\text{rate of particle deposition with repulsive electrostatic forces}}{\text{rate of particle deposition without electrostatic forces.}}$$

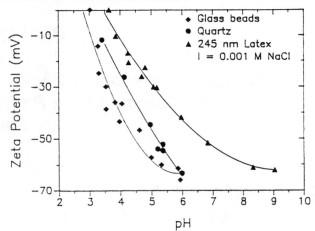

Figure 1. Zeta potential measurements of crushed glass and quartz particles and carboxyl latex spheres.

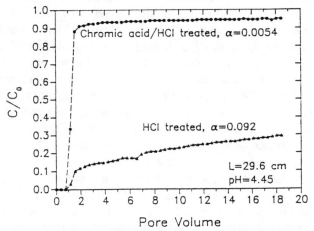

Figure 2. Comparison of breakthrough curves for 0.245 μ latex colloids and 250-300 μm soda-lime glass beads with different surface cleaning techniques.

The theoretical attachment efficiency, α_t, in the presence of an energy barrier, was determined by the procedure described by Elimelech and O'Melia (1990a). Repulsive electrical double-layer forces were estimated using zeta potentials computed from electrophoresis measurements of crushed quartz and glass particles and latex spheres (Figure 1). Experimental attachment efficiencies were determined from mass balance calculations (Yao et al. 1971).

Filtration experiments were performed under unfavorable conditions with 0.245-μm-diam carboxyl latex spheres (Interfacial Dynamics Corporation) in glass-wall columns packed with 250 to 300-µm-diam soda-lime glass beads (Cataphote, Inc.) or ultrapure quartz rains (Feldspar, Inc.) at porosities of approximately 0.39 and 0.48, respectively. The approach velocity was maintained at 0.030 cm/s and the ionic strength was adjusted to 0.001 M with NaCl. The pH within the column was allowed to equilibrate prior to filtration by passing the electrolyte solution without particles through the column until the influent and effluent pH values were within 0.01 unit. During the equilibration periods and filtration experiments, the effluent pH was continuously measured in a small flow-through cell. The influent concentration of the latex spheres was approximately 0.25 mg/L. Influent and effluent particle concentrations were determined continuously with a spectrophotometer equipped with a flow-through cell. Aggregation of the latex particles was monitored with the use of a submicron particle size analyzer.

Experimental attachment efficiencies were dependent on the collector media cleaning method. Figure 2 exhibits the breakthrough curves for glass-bead collectors subjected to either concd. HCl or concd. commercial chromic acid and concd. HCl. Chromic acid was selected to remove possible organic surface contaminants and was used after the glass beads were soaked in concd. HCl for 24 h, followed by thorough rinsing with deionized water (18 MΩ cm). Once the beads were exposed to chromic acid for 24 h, they were rinsed with deionized (DI) water, soaked again in concd. HCl, thoroughly rinsed with DI water, boiled 4 times in fresh 4 N HNO$_3$ for 1 h, batch-rinsed with DI water, and stored wet at 4°C. This treatment resulted in a decrease in the attachment efficiency of more than one order of magnitude at pH 4.45 and ionic strength of 0.001 M (NaCl). As shown in Figure 3, the removal of organic matter from the surface resulted in deposition behavior that was more consistent with theory, but significant discrepancies with theory are still evident at higher pH values.

Filtration experiments performed with quartz-grain collectors also exhibited a dependence on surface treatment as presented in Figure 4. Quartz that was sieved to obtain a uniform size distribution resulted in higher deposition rates than quartz that was sized by sedimentation techniques, even after exposure to concd. HCl for 24 h. The lowest attachment efficiencies were observed when the sedimented-sized quartz was calcined at 810°C for 8 h to combust possible organic surface contaminants. Quartz exposed to the chromic acid treatment employed with the glass beads exhibited deposition behavior that was identical with the calcined quartz. Attachment efficiencies using glass-bead collectors exposed to the chromic acid/HCl cleaning treatment were somewhat greater, but similar to the values measured with well-cleaned quartz collectors. The theoretical attachment efficiencies for the quartz and glass collectors are indistinguishable when calculated from the zeta potential measurements presented in Figure 1.

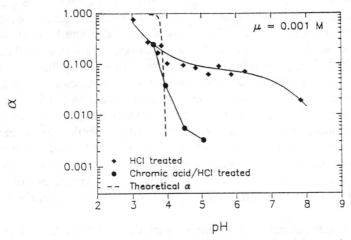

Figure 3. Effects of surface cleaning treatments and sizing techniques on the attachment efficiencies for latex particle/quartz-grain collector filtration experiments.

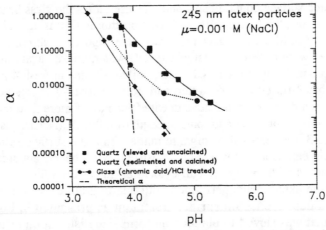

Figure 4. Effects of surface cleaning treatments and sizing techniques on the attachment efficiencies for latel particle/quartz-grain collector filtration experiments.

The results suggest that materials commonly used for filtration experiments possess surface contaminants that enhance particle deposition unless adequately cleaned. Particle deposition rates are significantly more consistent with theoretically calculated behavior once the collector surfaces are aggressively cleaned of contaminants, however, discrepancies still exist between experimental and predicted values.

REFERENCES

Bowen, B.D. and Epstein, N., Fine particle deposition in smooth parallel-plate channels, *J Colloid Interface Sci* 72:81-97, 1979.

Elimelech, M. and O'Melia, C.R., Effect of electrolyte type on the electrophoretic mobility of polystyrene latex colloids, *Colloids Surf* 44:165-78, 1990a.

Elimelech, M. and O'Melia, C.R., Effect of particle size on collision efficiency in the deposition of Brownian particles with electrostatic energy barriers, *Langmuir* 6:1153-63, 1990b.

Wishart, G.J., Deposition of latex particles on alumina fibers, *Colloids Surf* 1:313-34, 1980.

Rajagopalan, R. and Chu, R.Q., Dynamics of adsorption of colloidal particles in packed beds, *J Colloid Interface Sci* 86:299-317, 1982.

Tobiason, J.E., Chemical effects on the deposition of non-Brownian particles, *Colloids Surf* 439:53-77, 1989.

Yao, K., Habibian, M.T., and O'Melia, C.R., Water and waste water filtration: Concepts and applications, *Environ Sci Tech* 5:1105-12, 1971.

CHAPTER 45

Sampling, Characterization, and Transport Studies of Groundwater Colloids

G. Longworth and M. Ivanovich

The study of the effects of colloids on the subsurface transport of waste materials, either radionuclides or inactive chemical species, depends on the development of effective techniques for the field sampling of colloids without interference and their subsequent characterization. The work outlined here deals with field sampling using tangential flow ultrafiltration and characterization of the colloids in terms of physical, chemical, and actinide composition (Longworth and Ivanovich 1989). The emphasis has been on investigating the effect of groundwater colloids on radionuclide migration, initially by measuring the partition of natural series actinides between the colloid and solution phases, leading to studies of colloid migration. Many of the problems involved in sampling, storage and characterization are, of course, relevant to studies of the transport of toxic chemical and heavy metal contaminants in groundwater.

During sample collection, the existing geochemical equilibrium in the water/colloid/rock system must not be significantly altered. Contamination of the colloid population must be kept to a minimum (Degueldre et al. 1990). The volume of colloid sample required is small (10 to 10^2 mL) for physical and chemical characterization but the measurement of actinide activities by alpha spectrometry requires larger quantities (10^2 to 10^3 L).

Cross-flow ultrafiltration (Degueldre et al. 1990) provides a sample in a convenient form for enumeration using electron microscopy, but the physical configuration of the colloids may be altered by trapping them on a solid membrane.

During tangential flow ultrafiltration the colloids remain in an aqueous medium and a concentrated sample can be obtained. Control of the atmosphere within the ultrafiltration rig ensures that the possible creation of new colloids is minimized. The tangential nature of the filtration generally leads to only small adsorptive losses on the filter. However in the case of inorganic colloids these losses may still be a significant fraction of the total colloid population (Degueldre et al. 1990).

Elemental analysis of colloids is conveniently carried out using Inductively Coupled Plasma Emission Spectroscopy (ICP/ES), while colloid sizes above about 30 nm may be assessed using electron microscopy. Laser Induced Photoacoustic Spectroscopy has been used to determine colloid concentrations for sizes above about 1nm, while microelectrophoresis can be used to study colloid mobility.

In studies of groundwater colloids at a series of aquifers (slate, granite, glacial sand, and sandstone) concentrations of colloids >50nm in size in the range 10^9 to $10^{12}/L^{-1}$ were measured. For the silica colloids from a granite aquifer, a weight concentration of about 0.05 mg/L was determined.

The fraction of total actinide activity associated with the colloid phase for inorganic colloids is small, 0.04 to 0.3% (uranium) and 1 to 10% (thorium). The daughter/parent activity ratio $^{234}U/^{238}U$ indicates that uranium in the colloids is in chemical equilibrium with that in solution.

Corresponding measurements in organically rich groundwaters (DOC = 100 mg/L) from the sedimentary layers overlying a salt dome, indicate that here the bulk of actinide concentration is associated with the humic colloids. The thorium concentration (6×10^{-9} mol/L) is two orders of magnitude greater than the values found in inorganic groundwaters.

The values obtained for the activity ratio $^{234}U/^{238}U$ reveal that the uranium in the colloid phase is not in chemical equilibrium with that in the solution phase.

ICP measurements on the filtrates, following cross-flow ultrafiltration with filters having a range of pore sizes, show that trivalent and tetravalent metals are also strongly bound to the colloid phase. Most of the colloid population occurs in the smallest size range of 15 to 1.5 nm, which provides the largest surface area for potential interaction with radionuclides/impurity elements.

Similar measurements on saline waters of low organic content from the same site show <20% of total uranium or thorium associated with the colloid phase, in line with the earlier measurements in inorganic groundwaters.

To assess colloid mobility in the field two approaches are possible. In the first the colloid characteristics are measured in a set of groundwaters from a known flowpath in an aquifer, while in the second a colloid tracer test is carried out in a given aquifer.

As a preparation for such a tracer test, laboratory column experiments in glacial sand have been carried out, and have demonstrated that the sorption of monodisperse synthetic colloids of hematite and latex increases with increasing ionic strength and with decreasing colloid charge. These studies will provide the basis for a field tracer measurement of colloid transport, which will yield input parameter values for a colloid transport model.

Thus there is evidence that humic colloids may be capable of transporting a significant proportion of mobile actinides. This work confirms that modelling of radionuclide transport requires the rock/water system to be treated as a three phase system, comprising the rock substrate, the groundwater/pore water and the colloid phase.

REFERENCES

Degueldre, C., Longworth, G., Moulin, V., and Vilks, P., Grimsel colloid exercise, CEC report EUR 12660 EN, 1990.

Longworth, G. and Ivanovich, M., The sampling and characterization of natural groundwater colloids: studies in aquifers in slate, granite and glacial sand, Nirex report NSS/R165, 1989.

CHAPTER 46

Conformational and Size-Change Studies on Model Organic Polyelectrolytes

N.A. Marley and J.S. Gaffney

If colloidal or macromolecular materials are to be used successfully in mixed waste cleanup in complex subsurface aqueous media, we must understand how their chemical and physical natures change as they interact with the geochemical environment and the waste media. Of particular interest are the natural organic complexing agents observed in groundwaters. These humic and fulvic acid macromolecules are known to be strong complexing agents for trace metals and actinides and have been implicated in the transport of actinides and other radionuclide materials. Humic and fulvic acids are known to have complex chemical structures that can interact with both hydrophilic and hydrophobic groundwater pollutants and enhance their transport in subsurface systems (McCarthy and Zachara 1989).

In the case of toxic metal and actinide binding to humics, carboxylate functional groups have been identified as important complexing features. Conformational effects resulting from changing pH have also been proposed as a means of explaining the complicated kinetics of metal binding to humics (Cacheris and Choppin 1989). Past attempts to simplify and characterize humic and fulvic acid interactions in aqueous solutions, have used model polyelectrolytes with known carboxylate structures (Choppin and Cacheris 1990).

In this work, the polyelectrolytes polymaleic acid (PMA) and polyacrylic acid (PAA) have been chosen for study because they have strong binding constants for many toxic metals and radionuclides (Choppin and Cacheris 1990). Fourier transform infrared spectroscopy using internal reflectance methods has been used to examine the conformational changes of these compounds as a function of pH in aqueous solutions. Using silicon and germanium rods, we have obtained aqueous infrared data with CIRCLE

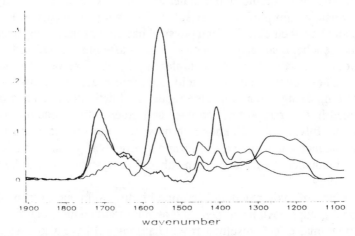

Figure 1. Conformational changes of Polymaleic Acid with varying pH.

Figure 2. FTIR spectra of Polyacrylic Acid in aqueous solution, pH = 3.0, 4.5, 7.0.

cell techniques. PAA and PMA Raman spectra and turbidity data for aqueous solutions as a function of pH have also been obtained (Marley unpublished data). The data support three basic conformational structures for the model polyelectrolytes that are summarized in Figure 1. At lower pH, the polycarboxylic acids were totally protonated and tended to take on a hydrophobic colloidal conformation in solution. As the pH was adjusted near the first pKa of the acids (around pH 4.5) the polymeric organic diacids were observed to form internal hydrogen bonds that led to the formation of a highly ordered crystalline form. At higher pH the totally deprotonated carboxylates were observed to be dissolved. Infrared spectra obtained for these solutions as a function of pH (Figure 2). Figure 2 supports and complements the observations of the laser Raman studies. The protonated acid carbonyl stretch is observed at 1710 cm^{-1}, while the carboxylate is at 1555 cm^{-1}. Light-scattering studies also support the formation of a colloidal structure at low pH, whereas at high pH the polyelectrolytes stay in solution.

Preliminary studies on humic and fulvic acids indicated that the presence of metal ions in the solution and the subsequent complexation to the carboxylate groups in the natural polyelectrolytes can be observed directly with these methods. Both pH and metal ion content and other geochemical parameters will affect the conformations of humic and fulvic acids in groundwaters. This work strongly indicates that these conformational changes can affect the capacities of humic and fulvic acids to bind both hydrophobic and hydrophilic contaminants and their subsequent transport in surface environments.

The model studies using PMA and PAA and the preliminary investigations of humic and fulvic materials indicate that the binding and solubility behavior of organic colloidal materials and macromolecules must be understood on a physicochemical level if we are to use synthetic or natural polyelectrolytes in cleanup procedures.

REFERENCES

Cacheris, W.P. and Choppin, G.R., Dissociation kinetics of thorium-humate complex, *Radiochim Acta* 42:185-90, 1987.

Choppin, G.R. and Cacheris, W., Kinetics of dissociation of thorium (IV) bound to PMA and PMVEMA, *Inorg Chem* 29:1370-74, 1990.

McCarthy, J.F. and Zachara, J.M., Subsurface transport of contaminants: Mobile colloids in the subsurface environment may alter the transport of contaminants, *Environ Sci Technol* 23:496-502, 1989.

CHAPTER 47

On the Importance of Inorganic and Organic Colloids in Environmental Systems Relative to Nuclear Waste Disposal

V. Moulin, T. Dellis, C. Moulin, J.C. Petit,
D. Stammose, M. Theyssier, M.T. Tran, M.G. Della,
C.J. Dran, and J.D.F. Ramsay

Migration phenomena of radionuclides in geological systems are of great interest for the safety assessment of nuclear waste disposal. The mobility of trace radioelements in natural groundwaters is governed by complex physicochemical interactions that depend largely on the characteristics of the radioelements, the aquifer, and the rock matrices. Among others, the presence of colloids defined as particles of 1 to 1000 nm in size may affect the behavior of radioelements by different processes such as sorption, complexation, dissolution-precipitation (Ramsay 1988; Choppin 1988; McCarthy and Zachara 1989; Moulin and Ouzonnian unpublished data). These colloids may be organic (humic substances) and inorganic (silica, oxides, and clays particles). The identification of colloids present in natural waters (determination of their concentration, composition, and properties) and the understanding of the reactions involved between colloids and radioelements appear to be of great importance in predicting the speciation of radioelements in aquifer systems. Research has been carried out in these fields to determine the importance of colloids in environmental systems in relation with nuclear waste disposal.

The occurrence of colloids (inorganic and organic) has been studied in various waters, mainly granitic groundwaters: the Grimsel groundwater (Grimsel/Switzerland, a crystalline rock) selected from an intercomparison exercise carried out in the Commission of the European Communities and the Fanay-Augeres water (Massif-Central/France, a U-mine) selected as a

297

Table 1. Characteristics of the Fanay-Augères humic substances (HA and FA stand for humic acids and fulvic acids, respectively).

| | Fanay-Augères | |
	HA	FA
Water sampling	280 m	
Water analysis		
pH	6.0	
HCO$_3$	24	
SO$_4$	10	
Cl	6.5	
Na	8.5	
K	1	
Mg	1	
Ca	4	
Fe	<0.01	
Datation (years)		
tritium (on water)	16	
organic	625	-
Elemental composition		
% C	46.2	49.2
% H	6.3	4.8
% O	30.6	44.7
% N	8.0	1.4
% ash	<1	3.0
Mineral composition (ppm)		
Al	340	20
Fe	670	40
Ca	5080	2200
Mg	2200	10
U	7760	30
Acidity (meq/g)	3.4	5.7
Molecular weight (Mw)	5000*	5000*
Size (nm)	50%	90%
	3-25 nm	<1.5 nm

*Shodex column (tris buffer pH 7)

Table 2. Characteristics of the Grimsel colloids (colloid concentration obtained on the Grimsel water by SEM counting on filters prepared on site).

Pore size (nm)	[Colloid] part/l	Size range (nm)
3	$4.5 \cdot 10^9$	70-1000
15	$3.8 \cdot 10^9$	70-1000
15	$4.4 \cdot 10^9$	70-1000
100	$6.0 \cdot 10^8$	100-1000
220	$1.8 \cdot 10^8$	220-1000
450	$2.6 \cdot 10^8$	450-1000

Table 3. Interaction constants (K in mol^{-1}) and complexing capacities (W in mmol/g) obtained for Aldrich and Fanay-Augères humic substances by spectrophotometry (SP), size-exclusion chromatography (SEC), and laser-induced time-resolved spectrofluorometry (LITRS) at a ionic strength of 0.1 *M*.

Method	Conditions*	pH	Aldrich humic acids		Fanay-Augères Humic acids		Fanay-Augères Fulvic acids	
			W	log K	W	log K	W	log K
SP	0-100 mg/l	4.65	0.96	7.0	0.3	7.0	0.45	6.5
(Am)	3 10-5 M			(4.0)**		(3.5)		(3.2)
SEC	0-500 mg/l	5	-	(4.8)	-	(4.6)	-	(4.2)
(Am)	10-7 M							
LITRS	0-10 mg/l							
(Cm)	5 10-8 M	5	0.9	8.7			0.03	9.3
				(5.6)				(4.8)
	0-50 mg/l							
	10-6 M	5	0.8	8.0			0.07	7.8
				(4.9)				(3.7)

* ligand concentration (in mg/l) and cation concentration (in M) ** K expressed in l/g

water representative of a granitic formation. Various methods such as transversal ultrafiltration for colloids (Degueldre et al. 1989) and ion-exchange resins for humic substances (Dellis and Moulin 1989) have been used for the concentration and extraction of these colloids. Their characterization by means of various techniques (mainly chemical and size analysis) has been carried out to determine their principal properties. The main characteristics of humic substances isolated from the Fanay-Augeres aquifer and Grimsel colloids are listed in Tables 1 and 2, respectively.

The interactions of these colloidal entities with radionuclides have also been studied. In the case of humic substances, different analytical techniques have been developed for studying complex formation with trivalent elements (Am and Cm). These techniques include size-exclusion chromatography based on the separation of the macromolecular complex from the free metal on a porous gel (Lesourd-Moulin 1986; Moulin et al. unpublished data), spectrophotometry (Moulin et al. 1987), and laser-induced time-resolved spectrofluorometry (Moulin et al. unpublished data), based on the specific properties of the elements (absorbance or fluorescence) and their titration by humic substances. These methods enable determination of complexing capacities of humic substances toward the cation and calculation of the conditional interaction constants, assuming the formation of complexes with a 1:1 stoichiometry. Table 3 summarizes the main results obtained.

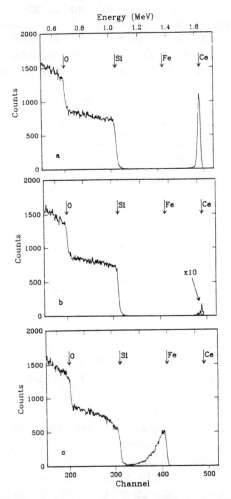

Figure 1. RBS spectra of silica samples after 7 days of contact with ceria colloids (a), silica colloids coated with ceria (b) and hematite colloids (c).

Because of the strong affinities of organic and inorganic colloids for mineral surfaces, their effect on radioelement retention has been investigated to determine whether they could modify the behavior radioelements (increase or decrease their mobility). Sorption of colloids (Ce, U, and Th) and pseudocolloids (association of Ce, Th, and U with colloidal silica or iron oxides) onto mineral surfaces (silica, mica, and iron oxides) has been studied by using Rutherford Backscattering Spectrometry (RBS), which gives information on the amount and depth distribution of retained colloids (Billon et al. 1989). It has been shown that colloid and pseudocolloid

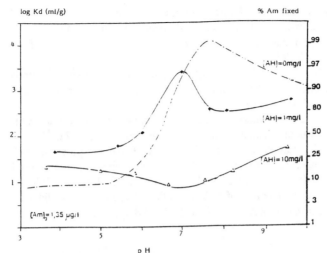

Figure 2. Am retention onto alpha alumina as a function of pH and humic acid concentration (ionic strength of 0.1 *M*).

retention depends largely on the charge; charge density and structure of the interface, and the charge, size and structure of colloids (Figure 1).

The influence of humic substances on Am retention onto silica and alumina has also been studied (Moulin and Stammose 1988; Moulin et al. 1989), showing that humic acids control the behavior of the cation [i.e., organic film is formed at the solution, and retention is decreased as a result of organic complexation in solution (Figure 2)].

REFERENCES

Billon, A., Caceci, M., Della, M.G., Dran, J.C., Moulin, V., Petit, J.C., Ramsay, J.D.F., Russel, P.J., and Theyssier, M., The role of colloids in the transport of radionuclides in geological formations, AERER-13691, 1989.

Choppin, G.R., Humics and radionuclide migration, *Radiochim Acta* 44/45:23-28, 1988.

Degueldre, C., Longworth, G., Moulin, V., and Vilks, P., Grimsel colloid exercise, CEC Report EUR 12660 EN, 1989.

Dellis, T. and Moulin, V., Isolation and characterization of natural colloids, particularly humic substances, present in a groundwater, in Miles, D.L., (ed.), Water-Rock Interaction WRI-6, Balkema, 1989, pp 197-201.

Lesourd-Moulin, V., Humic acids and their interactions with metallic elements Cu(II), Eu(III), Th(IV) and U(VI), Report CEA-R-5354, 1986.

McCarthy, J.F. and Zachara, J.M., Subsurface transport of contaminants, *Environ Sci Technol* 23:496-502, 1989.

Moulin, C., Decambox, P., Mauchien, P., Moulin, V., and Theyssier, M., On the use of laser-induced time-resolved spectrofluorometry for interaction studies between organic matter and actinides: Application to curium, *Radiochim Act* 52(53):119-125, 1991.

Moulin, V., Robouch, P., Vitorge, P., and Allard, B., Spectrophotometric study of the interaction between americium (III) and humic materials, *Inorg Chim Acta* 140:303-06, 1987.

Moulin, V. and Stammose, D., Effect of humic substances on americium (III) retention onto silica, *Mat Res Soc Symp Proc* 127:723-27, 1988.

Moulin, V., Stammose, D., and Tran, M.T., Chemical processes at the mineral oxide-water interface: system Al_2O_3-organic matter-Am(III), Miles, D.L., (ed.), in Water-Rock Interaction WRI-6, Balkema, 1989, pp 505-509.

Ramsay, J.D.F., The role of colloids in the release of radionuclides from nuclear wastes, *Radiochim Acta* 44/45:165-70, 1988.

CHAPTER 48

Synchrotron Radiation Studies of the Sorption of Chromium Colloids

Dale L. Perry

Under geologic conditions, chromium can exist in both the hexavalent and trivalent oxidation states. Previous work in our laboratory (Perry et al. 1984a) has shown that the dichromate ion $Cr_2O_7^{2-}$ a major chemical species in some Department of Energy toxic waste sites, can undergo a two-step process in its interaction with reducing geologic substrates such as sulfides.

Step I: $Cr^{VI}_2O_7^{2-}$ reduced to $Cr^{III}_2O_3 \cdot xH_2O$ (CrOOH)
Step II: $Cr^{VI}_2O_7^{2-}$ monomerized to $Cr^{VI}O_4^{2-}$ (PbCrO$_4$)

Such reduction-oxidation systems produce a layering and mixing (Perry 1990) of colloid reaction products, the distribution and morphology of the products being dependent on the precise chemical state of the reducing sulfide surface. Reduction of the Cr(VI) to Cr(III) species and its deposition on the surface results in a passivation of the reducing sulfide followed by the subsequent chemisorption of the chromate species. The use of scanning electron microscopy studies confirms that the resulting colloid-covered surface can only be modeled as a heterogeneous array of chemical components available for undergoing subsequent dissolution by ground-water.

This surface heterogeneity is manifested both chemically and morphologically. The dichromate ion, with a formal electronic configuration of [Ar] $3d^0$, undergoes a three-electron reduction to the paramagnetic [Ar] $3d^3$ configuration in forming hydrated Cr_2O_3. In the presence of a carbon dioxide/carbonate source, a second phase—a hydrated oxide/carbonate—is formed. This secondary reaction is analogous to that previously reported to occur in both hydrated oxides of lead (Perry and Taylor 1984) and uranium (Perry et al. 1984).

The current research addresses the use of synchrotron radiation for extracting X-ray fluorescence microprobe data for chromium colloids generated on the surface of natural galena, PbS. Lateral elemental profiles are presented, and the X-ray fluorescence data are compared with data from previous spectroscopic studies. Various intermediates in the reaction pathway to the final products will be discussed, along with the implications they present for approaches to models.

ACKNOWLEDGMENTS

This work was supported by the U.S. Department of Energy under Contract NO. DE-AC03-76SF00098 and the Office of Energy Research, Office of Basic Energy Sciences, Engineering and Geosciences Division.

REFERENCES

Perry, D.L., Chapter 1, Instrumental Surface Analysis of Geologic Material, VCH Publishers, 1990.

Perry, D.L. and Taylor, J.A., An X-ray photoelectron and electron energy loss study of the oxidation of lead, *J Vac Sci Technol A* 2:771-74, 1984.

Perry, D.L., Tsao, L., and Taylor, J.A., The galena/dichromate solution interaction and the nature of the chromium(III) species, *Inorg Chim Acta* 85:157-60, 1984a.

Perry, D.L., Tsao, L., and Brittain, H.G., Photophysical studies of uranyl complexes, 4. X-ray photoelectron and luminescence studies of hydrolyzed uranyl salts, *Inorg Chem* 23:1232-37, 1984b.

CHAPTER 49

Laboratory Studies on the Stability and Transport of Inorganic Colloids Through Natural Aquifer Material

R.W. Puls and R.M. Powell

Colloids are generally defined as inorganic or organic particles of >10 µm diam. Colloidal material may be released from the soil or geologic matrix and transported large distances, given favorable hydrological and geochemical conditions. Once released, the primary factors controlling colloidal transport in subsurface systems are colloidal stability, flow rate, and the nature of the solid matrix through which groundwater flows.

In addition to having a high surface area per unit mass, colloids such as clay minerals and iron oxides are also extremely reactive sorbents for metals and other contaminants. If mobile in subsurface systems, these colloids can cause the migration of sorbed contaminants over much larger distances than current transport models would predict. Column experiments were run to determine the effects of pH, flow rate, ionic strength, electrolyte composition, particle size, and particle concentration on colloidal stability and transport.

Spherical radiolabeled iron oxide (α-Fe_2O_3) colloids ranging from 100 to 800 nm in diameter were synthesized according to Matijevic and Scheiner (1978). Stability, in terms of coagulation, was monitored by using laser light scattering with photon correlation spectroscopy. The colloids were stable in 0.005 M $NaClO_4$ and 0.005 M NaCl over the pH regions 2.0 to 6.5 and 9.7 to 11.0. From pH 6.5 to 7.6, the colloids were extremely unstable and from pH 7.6 to 9.7 they were quasi-stable [i.e., kinetics of coagulation were slow (several hours)]. The estimated point of zero charge from titration was pH 7.3 to 7.6. Some runs were conducted as low as pH 7.6, where the particles were relatively unstable; however,

Table 1. Summary of selected column runs

Size (nm)	pH	Flow Rate (mL/min)	Part.Conc. (mg/L)	Ionic strength	Anion eluted	%C_0
200	3.9	2.0	10	0.005	Cl^-	0
125	8.9	1.2	10	0.005	Cl^-	54
150	8.1	1.2	5	<0.001	ClO_4^-	57
150	8.9	1.2	5	0.03	SO_4^{2-}	14
100	7.6	1.2	5	0.03	HPO_4^{2-}	99+
100	7.6	1.2	5	0.03	$HAsO_4^{2-}$	97

colloidal size measurements in the effluent were performed and compared with influent measurements.

Aquifer material collected from a mining site near Globe, Arizona, was used for the column matrix. This site is a heterogeneous sand and gravel alluvial aquifer with estimated hydraulic conductivities ranging from 10 to 200 m/day. Material passing a 2-mm sieve but retained on a 0.1 mm sieve was used to pack the glass columns 2.5 cm in diameter and adjustable in length. Column lengths used ranged from 2.5 to 5.1 cm. Flow rates used were comparable to estimated groundwater flow velocities at the site.

Relatively high concentrations of phosphate have been observed to enhance colloidal stability through charge reversal on positively charged particles (Liang and Morgan 1990). A shift in the pH_{zpc} or increase in the stability region where the colloids are negatively charged was also observed. The effect of several different anions on the stability and transport of the colloids was determined in this study. No transport of colloids was seen on the positive side of the pH_{zpc}, indicating electrostatic interaction with the net negatively charged matrix material. Significant differences were caused by the anionic composition of the background electrolyte (Table 1). Specific adsorption of some anions caused surface charge reversal and enhanced stability of the iron oxide colloids. As increase in the net negative charge on the colloid resulted in an increase in electrostatic repulsion between the mobile colloids and the column matrix material.

Breakthrough of the colloids in advance of tritiated water, a conservative tracer, was observed to some extent in most of the column experiments. As much as 99+% of the influent concentration of colloids (C_0) was observed in the effluent, depending on pH, anionic electrolyte composition, and particle size. Increasing size exclusion volume was observed with increasing particle size; however, these differences may have been the results of different column packing characteristics and/or different geomedia (i.e., core from different wells).

Arsenate is a ubiquitous contaminant, and its chemical behavior is similar to that of phosphate. Typically, iron minerals have a very large adsorption capacity for arsenic. Column experiments with 1 mg/L dissolved arsenate in 0.01 M $NaClO_4$ were compared with arsenate surface-saturated iron oxide colloidal transport. Dissolved arsenate was retarded 21 times that of the colloidal fraction. When the columns were flushed with deionized water (low ionic strength), there was a spike of colloidal arsenic; when flushed with phosphate, there was a spike of exchangeable arsenate. These spikes were approximately equal to the influent arsenate concentration.

The facilitated transport of metal and metalloid contaminants may occur in porous media by adsorption of the contaminants onto mobile colloidal iron oxide, given appropriate physicochemical conditions. Transport is highly dependent on colloidal stability and anionic composition of the background electrolyte. In all experiments, the iron colloids were transported faster than tritiated water, primarily because of size exclusion effects. Maximum breakthrough, as percentage of influent colloid concentration ($\%C_o$), was strongly correlated with major anion composition and particle size.

DISCLAIMER

Although the research described in this article has been funded wholly or in part by the United States Environmental Protection Agency, it has not been subjected to the agency's peer and administrative review. Thus, this report may not necessarily reflect the views of the agency, and no official endorsement may be inferred.

REFERENCES

Liang, L. and Morgan, J.J., Chemical aspects of iron oxide coagulation in water: laboratory studies and implications for natural systems, *Aquat Sci* 52(1):32-55, 1990.

Matijevic, E. and Scheiner, P., Ferric hydrous oxide sols. III, Preparation of uniform particles by hydrolysis of Fe(III), *Interface Sci* 63(3):509-24, 1978.

CHAPTER 50

Transport of TiO₂ Through Homogeneous and Structurally Heterogeneous Porous Media

J.E. Saiers, J.F. McCarthy, P.M. Jardine,
L. Liang, and G.M. Hornberger

Inorganic colloids may play an integral role in facilitating the transport of subsurface contaminants, such as radionuclides and trace metals. Development of a predictive theory of contaminant transport, then, will first require accurate characterization of colloidal movement through geologic materials. Many factors complicate prediction of colloidal transport in natural subsurface environments, including media heterogeneities (fractures, macropores, and channels) and variability in groundwater composition.

In this study, the transport of colloidal titanium dioxide (TiO_2) was investigated under varying conditions of groundwater pH and media characteristics. Specifically, laboratory columns were constructed in homogeneous and structurally heterogeneous fashions. Homogeneous columns consisted of clean quartz sand of mean diameter of either 0.3 or 0.9 mm. The structurally heterogeneous columns were constructed by embedding a tubule of the coarse-grained sand within a matrix of the fine-grained material.

An artificial groundwater containing 10^{-3} M NaCl was introduced to the top of these water-saturated columns at a constant rate. Once steady-state flow was achieved, the TiO_2 suspension, which was adjusted to pH = 4.0 or 8.5, was injected into the columns. Samples of effluent were collected at 0.1-pore-volume increments and analyzed for colloidal concentrations by spectrophotometry. The colloidal injection was terminated after colloidal input concentration equalled output concentration, and the artificial groundwater was reapplied to the columns until effluent concentrations returned to zero.

In both the fine- and coarse-grained homogeneous experiments, colloidal mobility is much greater at pH = 8.5 than at pH = 4.0. At pH = 8.5, colloidal recoveries are nearly 100% for both the fine and coarse media experiments. However, at pH = 4.0, only 26% and 47% of the total colloidal mass is recovered for the fine and coarse homogeneous experiments, respectively. The smaller recovery in the fine media experiments relative to the coarse (at pH = 4.0) is attributed to the comparatively larger number of adsorption sites resulting from the greater surface area on the fine-grained sand. In addition to percentage mass recoveries, the time to reach peak concentrations ($C/C_0 = 1$) illustrates the differences in colloidal transport between the two pH treatments. Specifically, at pH = 4.0 peak concentrations are reached at 30.2 pore volumes and 18.5 pore volumes for the fine media and coarse media, respectively. At pH = 8.5, however, time to peak occurs in advance of what is expected of a conservative tracer, suggesting the presence of size and/or charge exclusion effects.

The mechanism responsible for the reduced colloidal mobility at pH = 4.0 relative to pH = 8.5 is probably electrostatic adsorption. TiO_2 has a net positive surface charge at pH = 4.0, so it electrostatically adsorbs to the negatively charged quartz surfaces. This adsorption appears to be irreversible. At the basic pH, on the other hand, negative surface charges on both the quartz and the TiO_2 particles present adsorption resulting in the conservative transport of colloidal TiO_2.

As in the homogeneous experiments, colloidal transport through the structured, heterogeneous columns is much more efficient at pH = 8.5 than at pH = 4.0. For both pH treatments, however, analysis of the breakthrough curves indicate that transport is occurring in both the preferred flow path and the fine-grained matrix of the structured, heterogeneous columns. This two-domain flow is made apparent by the inflection in both the ascending and descending limb of the breakthrough curve for pH = 8.5 (see Figure 1). The data that make up the portion of the curve from the initiation of the experiment to approximately the second pore volume result from suspended particle delivery through only the preferred flow path. The interval between the second and third volume, on the other hand, represents colloidal transport through both the preferred flow path and the fine-grained matrix. After the injection is terminated, the decline in output concentrations to $C/C_0 = 0.4$ indicates that the preferred flow path is being flushed of the suspended material. Also during this time, it is likely that colloidal TiO_2 is diffusing from the matrix into the preferred flow path in response to the established concentration gradient between the two domains. This mass transport from the matrix to the preferred flow path, although relatively small, probably occurs from the moment that the injection is terminated until the end of the experiment. The region of the breakthrough curve from ~4.75 pore volumes to 5.5 pore volumes primarily represents elution of the matrix.

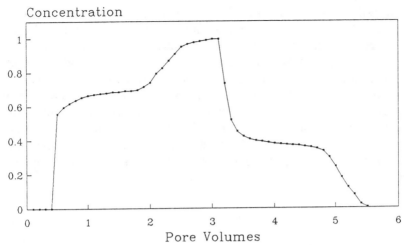

Concentration

Figure 1. Breakthrough curve for structured, heterogeneous media at pH = 8.5.

Pore Volumes

Although breakthrough data from the structured, heterogeneous experiments has not yet been modelled, progress has been made in modelling some of the results from the homogeneous experiments. Implementation of a convection-dispersion model resulted in accurate prediction of the observed data at the basic pH. Characterization of colloidal migration at pH = 4.0 will require that a parameter for irreversible adsorption be integrated into the current convection-dispersion model.

A dual-porosity expansion of the convection-dispersion model will be implemented to simulate colloidal transport through the structured, heterogeneous media. The conceptualization is that parallel flow occurs in both the preferred flow path and the fine-grained matrix. The flow in each zone can be described by the standard convection-dispersion equation. Besides parallel flow in each domain, an expression for mass exchange is incorporated into the model. It basically describes the movement of solute or suspended material between the two regions in response to a concentration gradient. For instance, if the preferred flow path is flushed with "fresh water," the colloidal suspension will diffuse from the matrix into the preferred flow path in response to the established concentration gradient.

Many of the model parameters for these proposed dual-porosity simulations can be measured independently of the experimental data. Through modelling the results of homogeneous experiments by convection-dispersion, the dispersion coefficient and adsorption parameter can be estimated for the preferred flow path and the fine-grained matrix. Similarly, by conducting falling head permeameter tests on homogeneous mixtures of each media type, the darcian velocity can be derived separately for each domain. The ability to obtain independent estimates of these model parameters leaves only the mass exchange coefficient to be fitted to the data.

It seems, then, that grain size, pH, and structural heterogeneity exert a strong influence over the transport characteristics of colloidal TiO_2. It is postulated that the fundamental concepts developed from these types of laboratory experiments and modelling exercises will aid in increasing current understanding of the manner in which inorganic colloids move through geologic materials. Ultimately, this knowledge can be applied to determining the transport behavior of contaminants in natural subsurface environments.

CHAPTER 51

Rescan - A Multi-Electrode Resistivity System for Monitoring Movement of Tracers

M.A. Sen, R.S. Ward, G.M. Williams, G.P. Wealthall,
P.D. Jackson, P.I. Meldrum, R.D. Ogilvy, and N. Bosworth

This paper describes a novel solute/pollution monitoring technique that was developed as part of an ongoing research project. This project is aimed at evaluating aquifer heterogeneity by undertaking a number of small-scale radial injection tracer tests within a larger domain under investigation (Williams et al. 1988).

For these tests a computer-controlled multielectrode resistivity system (RESCAN) has been developed that can make large numbers of measurements automatically (Jackson 1989). The system involves taking very few water samples and so avoids the high cost and problems of manually extracting and analyzing large numbers of groundwater samples.

Although this was the original purpose of the system, there are a large number of other potential applications including: saturated & unsaturated zone contamination studies; assessment of landfill liner integrity; saline intrusion in coastal aquifers; fracture identification and tracer tests in fractured rock; capillary zone and soil moisture monitoring; measuring vertical flows in boreholes; monitoring landslips and instability; void, cavity and shaft identification; discharge measurement by dilution gauging; monitoring saline & fresh water mixing in estuaries; delineating faults; determining three-dimensional glacial stratigraphy; identification of burial channels; and tomographic imaging of buried archeological features.

To test the system, a radial injection tracer test using sodium chloride was carried out with monitoring both by RESCAN and conventional multi-level samplers.

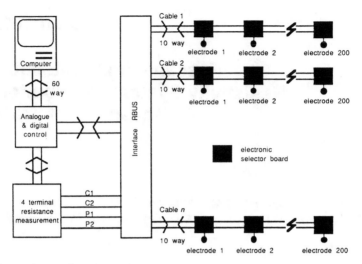

Figure 1. Computer-controlled multi-electrode resistivity system architecture.

The measurement of electrical resistivity is a well-tried and tested geophysical survey technique that has been applied in numerous geological contexts (Zohdy et al. 1984). The ability of rocks to conduct an electrical current is dependent primarily on the amount, salinity, and distribution of water within the rock. The presence of clays and conductive minerals also reduces the resistivity of the rock.

In conventional geophysical surveys, electrodes placed in the ground are connected manually or by mechanical switches to a current source or voltage measurement device, a process which is slow and laborious. With RESCAN, each electrode can be electronically selected and controlled by software either to pass current or measure potential. Thousands of electrodes can be attached to the system, giving unprecedented data density and target resolution. Figure 1 shows the architecture of the system.

In summary, the advantages of RESCAN are

- The ability to accommodate any number of electrodes.
- The ability to select electrodes and pass current in any direction, by means of software.
- The speed of measurement.
- The accuracy of measurement.
- The compatibility of the data storage architecture with existing graphics displays/processing systems.
- A wide number of applications using different probe designs.

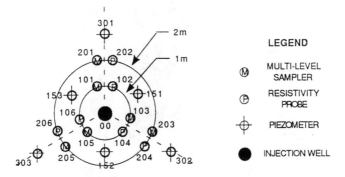

Figure 2.   Borehole and instrument array layout.

To fully demonstrate the resistivity technique in tracer tests, the results from a small-scale field experiment employing the described technology are presented. A radial injection tracer test was carried out in an unconfined river gravel aquifer underlain by clay at a depth of 6 m. Preliminary hydraulic testing of the array undertaken to investigate the average aquifer properties (Ward et al. 1990) gave average hydraulic conductivities over the site ranging from 46 to 146 m/day. The considerable variation is almost certainly caused by heterogeneity.

The instrument array for the tracer test consisted of the following components (Wealthall et al. 1990):

1. A central 6-in.-diam fully-screened injection/recharge well.
2. Six resistivity probes installed at 1 and 2 m radii from the central well along three limbs with a 120° angle between each limb. These were constructed from steel tubing with an outer casing of PVC water pipe for electrical insulation. Each had 80 insulated 5-mm diam brass electrodes spaced every 5 cm along their length. They were pushed into the ground using the hydraulic head-feed of a lorry-mounted drill rig.
3. Six multilevel samplers positioned adjacent to each of the resistivity probes, 30 cm away but at the same radii.
4. Six fully-screened piezometers, three at 1.5-m radius and three at 3.0-m radius.
5. A surface grid array of 140 electrodes. These were mild steel stakes 0.5 m long, hammered into the ground to a depth of 25 to 30 cm.

A plan of the completed instrument array is shown in Figure 2.

The test was carried out by first injecting fresh water into the central well at a rate of 4.5 L/min. until steady-state conditions had been reached. Then, 5500 ppm of chloride was injected to a total volume of 10.8 m³

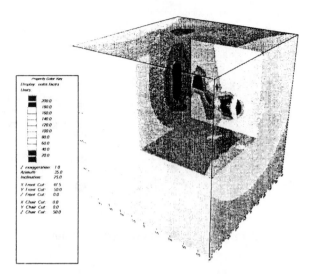

Figure 3.   3D geophysical model using synthetic data.

before switching back to fresh water injection. A very satisfactory square input pulse was achieved in this way.

During the experiment the tracer movement was monitored both by the RESCAN system and, for comparison, by extracting water samples from the multilevel samplers by means of a vacuum manifold board system (Hitchman 1988).

The resistivity probes were each used to give the equivalent of 77 very closely spaced multilevel sampler-type measurements by selecting groups of four adjacent electrodes as a Wenner array. The resistances measured by the probes are translated into pore water conductivities by using a formation factor for each measurement position. The formation factors were calculated from a background suite of water sample conductivity and RESCAN measurements using the formula

$$\text{Formation factor} \; = \; \frac{\text{Resistivity of formation}}{\text{Resistivity of pore water}}$$

If the rock matrix conductivity is comparable to that of the pore water, this calculation will not be completely accurate. For this reason it is desirable to make the measurements when the pore water has a high conductivity (i.e. when the tracer concentration is high). Unfortunately in this test the heterogeneity of the formation was such that the pore water conductivi-

102

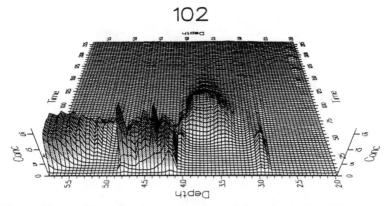

Figure 4.   Concentration vs depth and time for probe 102.

ty at the probes was not necessarily the same as that at the adjacent multi-level samplers.

Three-dimensional geophysical modelling techniques developed at British Geological Survey (BGS) enable the surface array measurements to be used to generate a three-dimensional picture of the plume movement. At this time, the surface array results have not yet been interpreted, but Figure 3 shows the sort of picture that can be generated using synthetic data.

With 77 breakthrough curves from each probe, it is not possible to present all the results here; however, Figure 4 shows a plot of concentration vs depth and time for probe 102. Three major zones of breakthrough can be seen. Some examples of individual breakthrough curves appear in Figure 5, along with the corresponding multilevel sampler breakthrough curves. The top graph shows one of the few cases in which the positions of the two curves coincide even though the magnitudes are different, probably because of rock matrix conduction leading to an underestimate in the RESCAN values. The others, however, show how different the curves could be even with measurement positions only 30 cm apart. This is because of the unanticipated degree of heterogeneity in the system, although it also seems likely that disturbance around the multilevel samplers, caused by the installation technique, played a part. The comparison also illustrates how easy it is to obtain very detailed curves by using RESCAN compared with the collection and processing of a much smaller number of water samples by using conventional multilevel sampler technology required a large amount of tedious work.

The viability of RESCAN, a novel tracer/contaminant monitoring device designed by the BGS, has been demonstrated. Its operation uses the well-known geophysical electrical surveying technique of resistivity mea-

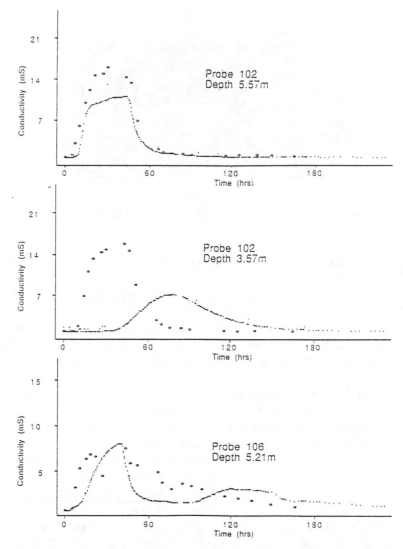

Figure 5.   Example breakthrough curves from RESCAN and multi-level samplers.

surement, and its novelty lies in the fact that it is much faster and more accurate than existing tools and is extremely versatile. Development of the system for use in radial injection tracer tests has also identified numerous other potential applications. Operation of RESCAN additionally requires much less manpower than standard monitoring techniques.

The results from the radial injection tracer test described show that the aquifer has a high degree of heterogeneity, which would almost certainly have been underestimated by a conventional monitoring system. In addition, once the results from the surface electrode array have been interpreted, an unprecedented three-dimensional picture of the tracer plume will be obtained.

ACKNOWLEDGMENTS

The authors are grateful to Stephen Hitchman of the British Geological Survey and Andy Dixon of the Institute of Hydrology for their contributions. This paper is published by permission of the Director of the British Geological Survey (Natural Environment Research Council).

REFERENCES

Hitchman, S.P., A collection manifold for multilevel ground-water sampling devices, *Groundwater* 26(3):348-49, 1988.

Jackson, P.D., Meldrum, P., and Williams, G.M., Principles of a computer controlled multi-electrode resistivity system for automatic data acquisition, WE/89/21, Fluid Processes Research Group, British Geological Survey, Nottingham, UK, 1989.

Ward, R.S., Sen, M.A., and Williams, G.M., Analysis of pumping test data from a small scale borehole array at Wallingford, WE/90/37, Fluid Processes Research Group, British Geological Survey, Nottingham, UK, 1990.

Wealthall, G.P., Ward, R.S., and Sen, M.A., Instrumentation of a radial injection tracer experiment using a multi-electrode resistivity system (RESCAN), WE/90/21, Fluid Processes Research Group, British Geological Survey, Nottingham, UK, 1990.

Williams, G.M., Noy, D.J., Jackson, P.D., and Mackay, R., Theoretical potential of radial injection tracer tests with 3-D pressure and solute monitoring, WE/88/26, Fluid Processes Research Group, British Geological Survey, Nottingham, UK, 1988.

Zohdy, A.A.R., Eaton, G.P., and Mabey, D.R., Application of surface geophysics to groundwater investigations, Chap. DI., in Techniques of Water-Resource Investigations of the United States Geological Survey, Book 2, 1984.

CHAPTER 52

Colloid Migration in a Confined Sand Aquifer

B. Smith, I. Harrison, J.J.W. Higgo, M.A. Sen,
P. Warwick, G. Wealthall, and G.M. Williams

A study of colloid migration encompassing both laboratory experimentation, fieldwork, and modelling is being performed by the Fluid Processes Research Group of the British Geological Survey. The objective of this work is to attempt to investigate the accuracy of predictive models of colloid mobility in a natural aquifer system, with particular emphasis on the effects of scale and local heterogeneity. The natural system under study is a well-characterized, confined, glacial sand aquifer (at Drigg in Cumbria, United Kingdom) in which extensive migration studies on soluble species have been performed and reported previously (Warwick et al. 1991; and Williams et al. 1991).

In the first phase of this project laboratory-column experiments were undertaken that were designed to (1) aid in the selection of suitable colloid(s) for use in field studies and (2) lead to an increased understanding of colloid behavior in the porous sand. The aquifer is predominantly quartz, with lesser quantities of orthoclase feldspar, kaolinite, mica, and chlorite. The particle size distribution was 95.2% sand (>63µm typically near spherical 100 µm in diameter), 0.2% silt (63 to 2 µm), 3.38% clay (2 to 0.5µm) and 1.20% clay (<0.5µm). Most of the experiments were carried out using polystyrene-latex fluorescent beads in sizes ranging from 0.055 to 0.600 µm. Preliminary experiments with ^{125}I-polyvinyl pyrolidone (PVP) have also begun. PVP has the advantage that it can be detected in the field by using the array of ranging gamma detectors already in place. PVP is available in molecular weights ranging from 5×10^3 to 1×10^6. Experimental Glass columns were slurry-packed with homogenized aquifer material and equilibrated with Drigg groundwater flowing at a linear velocity of 0.1 m/h. This packing procedure was found to yield consistent results with respect to porosity, and behavior of conservative tracer and

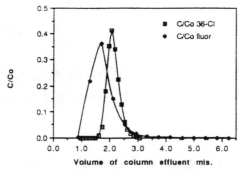

Figure 1.   Elution profile for a 7 cm column.

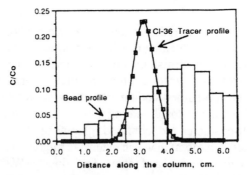

Figure 2.   Column profile showing transported material prior to elution.

colloids. The average effective porosity calculated from conservative tracers was 0.40. Two experiments were also carried out with "undisturbed cores" (i.e., cores that had been removed by boring the column into a bank of sand). The effective porosities of these were 0.41 and 0.37, respectively.

After calibration of the column with ^{36}Cl, pulse inputs of groundwater containing a high colloid concentration were injected into the flowing groundwater via a loop injection system, and the colloid content of the eluent was monitored. In some experiments a filter was placed at the end of the column. This had the effect of collecting the colloids and also preventing any "fines" escaping from the column. At the end of the run, the sand was extruded from the column and sectioned, and the colloid content of each slice was determined. Sand containing ^{125}I was counted in an automatic gamma counter. Sand containing fluorescent beads was shaken with acetone and filtered, and the fluorescence was measured.

Some of the profiles obtained are illustrated in Figures 1 and 2. Figure 1 shows the elution profiles of ^{36}Cl and 0.055-μm beads from a 6.7-cm

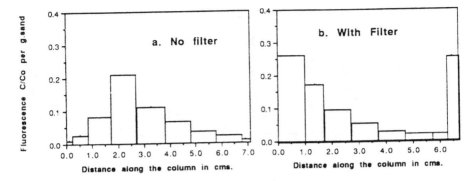

Figure 3.   (a) Column profile no end filter. (b) column profile with end filter.

column of repacked sand. The beads moved ahead of the ^{36}Cl, calculated Rf = 0.82 presumably because of size exclusion, but the peak was broader (dispersivity 9×10^{-4} m compared with 2×10^{-4} m). Figure 1a shows the profiles on an identical column of sand after only 0.52 column volumes had passed through (the ^{36}Cl profile was calculated from the previous profile). This clearly illustrates the faster travel time and greater dispersivity.

These experiments were carried out on short (<7-cm) columns, and about 80% of the beads were eluted. The remaining 20% were distributed along the length of the column. When the column length was increased to 18 cm, only 14% was eluted and the remaining 86% was spread along the column. It appears that a proportion of the beads was trapped, possibly by a filtration mechanism. In addition, longer term, experiments are under way in order to determine whether these beads are eventually released and over what time scale.

The behavior of iodinated PVP (38,000 Daltons) was similar to that of the 0.055-μm beads (i.e., it was eluted ahead of the conservative tracer and a proportion was distributed along the length of the column).

Figures 3a and 3b show the effect of placing a filter at the end of the column on the distribution of 0.510-μm beads along the column after about 30 column volumes of groundwater had passed through. In the absence of the filter, the beads seem to move along the column more readily than when a filter at the end prevents the escape of beads and fines. A possible explanation for this behavior (illustrated in Figure 4) is that channels form behind fines that are permitted to escape from the column and the beads then move down these channels.

Additional observations made during the course of the work were:

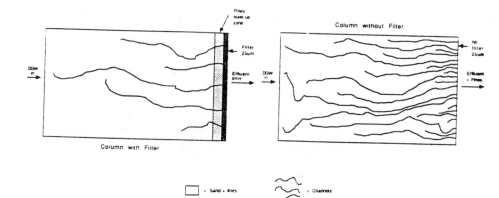

Figure 4.   Diagrammatic representation of the effect of a filter on the internal structure of a column.

1. In the absence of "fines" there was no filtration effect and even the largest beads moved through the sand columns without loss.
2. The mobility of the microspheres increased with decreasing colloid size.
3. The "plain microspheres" were more mobile than the carboxylated microspheres.
4. The behavior in the "undisturbed" cores was similar to that in the repacked cores although in one there was evidence of channelling.
5. Measurements of distribution ratios indicate that sorption of the microspheres on either the granular sand or fines is not the principal retention mechanism.
6. The experiments performed using packed columns with fines probably represent a "worst case" scenario in which only the most mobile-colloids migrate. Colloids such as the 0.055-μm noncarboxylated microspheres and PVP that have been shown to migrate under these conditions are therefore good candidates for use in field trials.

ACKNOWLEDGMENT

The authors would like to acknowledge the United Kingdom Department of Environment for funding this work.

CHAPTER 53

Preliminary Evidence for In Situ Enhanced Phenanthrene Transport in Column Studies

Thomas B. Stauffer and Leonard W. Lion

The main objective of this research was to demonstrate enhanced transport of phenanthrene in column studies. Problems with phenanthrene sorption to polymeric surfaces, such as Teflon®, were major obstacles. Both dynamic column and static batch sorption methods were developed to overcome phenanthrene sorption onto polymeric surfaces. Batch sorption isotherms were used to test a wide variety of minerals, particles, and chemicals for possible use in enhancing phenanthrene transport. During these experiments it was noted that changing the ionic strength of the aqueous solution used in transport studies produced a yellowish supernatant solution above the aquifer material. Column experiments showed differences in phenanthrene retardation when ionic strength was varied.

Distillate fuels are composed of a wide range of hydrocarbon chemicals. Chemical components range from short-chain aliphatic hydrocarbons to polycyclic aromatic hydrocarbons (PAH). In general, the aliphatic and low-molecular-weight chemical species are rapidly transported in low organic carbon content groundwater aquifers. Because these compounds move rather rapidly, remedial actions relying on pump-and-treat strategies are generally applicable for this class of chemical compound. However, distillate fuels, especially jet fuels contain significant amounts of highly substituted benzene rings and PAH. These intermediate-molecular-weight species are transported less rapidly in groundwater and, consequently, are not amenable to pump and treat remedial actions. Enhancing the transport of the substituted benzene rings and PAH may make it practical to use common pump and treat cleanup methods for fuel contaminated groundwater.

Phenanthrene, a PAH found in jet fuel, was selected for enhanced transport column studies. Because phenanthrene is very slightly water soluble, it sorbs relatively well to aquifer material; thus it was a good candidate for use in studies directed toward increasing PAH mobility. Mixed-batch sorption isotherms were measured on an aquifer material collected from the saturated zone at Columbus Air Force Base, Mississippi. During these experiments, it was noted that significant amounts of the PAH were sorbing onto the organic polymeric materials (Teflon®) commonly used to seal the bottles used in isotherm determination. An all-glass-sealed-ampoule batch sorption isotherm measurement technique was developed for obtaining sorption coefficients of phenanthrene on a low-carbon-content aquifer material. Because these same polymeric materials are found on commercial laboratory column end-fittings and seals, a custom designed all-glass column was constructed for use in the transport studies (Lion et al. 1990).

Column transport studies were run using C^{14} labelled phenanthrene as the retained solute and chloride as the nonretained tracer. A system consisting of two high-performance liquid chromatography (HPLC) pumps, a six-port valve, a glass column, and a fraction collector was used to measure the column breakthrough curves. Chloride concentrations were measured by chemical titration, and phenanthrene concentrations were measured by liquid scintillation counting. Phenanthrene retardation coefficients were calculated by first-moment analysis of the breakthrough curve.

Colloidal solids and dissolved macromolecules in soil pore water can potentially act as carriers for sorbed PAHs. Two conditions must be met to increase the PAH mobility: (1) the carrier must be able to bind PAHs to an appreciable extent and (2) the carrier must have a relatively greater mobility than the PAH molecule it binds. Most studies of potential carriers have focused exclusively on either their binding properties or their mobility and have not coupled these properties. Humic type macromolecules can bind hydrophobic pollutants (McCarthy and Jimenez 1985), but little evidence is present to show the extent to which humic-bound molecules are mobile.

Batch sorption isotherm methods were used to test a wide variety of minerals, particles, and chemicals to evaluate their potential for binding phenanthrene. With the exception of the Latex spheres, all colloidal solid suspension concentrations were 100 mg/L in 0.005 M $CaSO_4$ plus 0.02% NaN_3 electrolyte. Latex spheres were from Dow Chemical, supplied as a concentrated suspension, and were added to the electrolyte solution until the solution was turbid to the eye. None of the colloidal solids shown in Figure 1, except for the latex spheres, produced a stable suspension. Polystyrene and the latex spheres provided particles of known molecular weight and size.

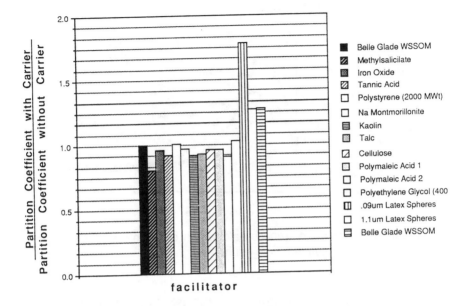

Figure 1. Phenanthrene batch partition coefficients in the presence of carriers relative to that observed in their absence.

With the exception of the Belle Glade water-soluble soil organic material (WSSOM), all dissolved organic solutions were 100 mg/L in the electrolyte solution discussed in the preceding. The WSSOM solution contained carbon at levels of ~80 mg/L after centrifugation at 1400 G and filtration through a Whatman GFC filter. Tannic acid was selected for use because its acidic functional groups make it likely to be ionized and mobile in soil. Polymaleic acid has been suggested as a synthetic analog for soil fulvic acids (Magee et al. 1989). WSSOM extracted from the Belle Glade muck was soluble or formed a stable suspension in the electrolyte solution.

Results for batch experiments in which phenanthrene sorption onto a graded sand was compared for aqueous solutions with and without selected characters are illustrated in Figure 1. The results indicate that methylsalicilate was the only material capable of substantially reducing the phenanthrene batch partition coefficient. Latex spheres and Belle Glade WSSOM actually acted to increase phenanthrene partitioning. Filtration of the WSSOM through a 0.45-μm filter eliminated its effect on the phenanthrene K_d, indicating the increased removal was caused by >0.45-μm particles.

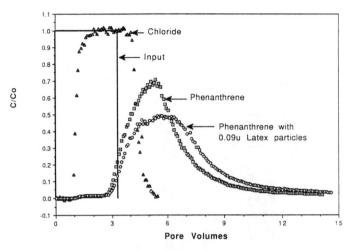

Figure 2.   Phenanthrene breakthrough curve in the presence and absence of latex spheres.

In batch experiments centrifugation was used to differentiate between sorbed and "dissolved" phenanthrene. Conceivably, colloidal solids (such as the Latex spheres) may be removed in this separation but may be stable in soil pore water. This is a limitation of the batch procedure to screen carriers for their ability to facilitate transport. In the case of the graded sand, adequate solid/solution separation could be achieved by simple sedimentation without centrifugation. Comparison of these two separation procedures revealed no effect on the observed phenanthrene K_d. Phenanthrene partitioning was still enhanced in the presence of the latex spheres when sedimentation was used to separate the solid phase. This result indicates that the increased phenanthrene removal occurred through binding of the latex particles to the sand. This result was confirmed by performing a column experiment on the sand in which the phenanthrene pulse contained 0.09-μm latex spheres. The resulting phenanthrene breakthrough curve is shown in Figure 2 in comparison to phenanthrene breakthrough in the absence of the microspheres. The column retardation value (calculated from the first temporal moment) for the sand increased to 5.0 in the presence of the latex spheres vs 4.1 in their absence. This corresponds to a 27% increase in K_d which is in qualitative agreement with the 78% increase observed in the batch experiments.

The above results indicate that batch experiments can indeed be used to deduce the influence of various carriers of interest. More important, the results show that PAH mobility may be decreased in low-carbon aquifers if colloidal solids are introduced that strongly bind to the aquifer matrix and also bind PAHs. Thus, it is possible to "defacilitate" as well as to

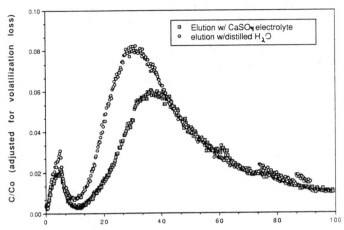

Figure 3. Phenanthrene breakthrough curves with electrolyte and in distilled water.

facilitate the transport of hydrophobic pollutants, depending on the nature of the pollutant, the "carrier," and the aquifer material.

During batch partition experiments it was noted that distilled water caused a yellowish-orange colored supernatant solution (possibly colloidal iron) over the Columbus aquifer material. Phenanthrene sorption isotherms with background electrolyte produced a K_d of 18.7, whereas the same material run in distilled water yielded a K_d of 14.6. Column experiments with and without background electrolyte showed differences in retardation vales from 146 to 121 calculated from two-site kinetic model fits (Parker and van Genuchten 1984). Breakthrough curves for phenanthrene on the aquifer material in electrolyte and in distilled water show early breakthrough for the distilled water relative to the electrolyte (Figure 3). Work continues to observe effluent particles, if any, and geochemical characteristics of the column effluent.

REFERENCES

Lion, L.W., Stauffer, T.B., and MacIntyre, W.G., Sorption of hydrophobic compounds on aquifer materials: analysis methods and the effect of organic carbon, *J Contam Hydrol* 5:215-34, 1990.

Magee, B., Lemley, A., and Lion, L.W., The effect of water-soluble organic material on the transport of phenanthrene in soil, Kostecki, P. and Calabrese, E., (eds.), Proceedings from the Third Conference on the Environmental and

Public Health Effects of Soils Contaminated with Petroleum Products, Lewis Publishers, Inc., Chelsea, MI, 1989.

McCarthy, J.F. and Jimenez, B.K., Interactions between polycyclic aromatic hydrocarbons and dissolved humic material: binding and dissociation, *Environ Sci Technol* 19:1072-76, 1985.

Parker, J.C., van Genuchten, M.T., Determining transport parameters from laboratory and field tracer experiments, Virginia Agricultural Experimental Station, Bulletin 84-3, Virginia Polytechnic Institute and State University Blacksburg, VA, 1984.

CHAPTER 54

Studies of Colloids and Suspended Particles in the Groundwaters of the Cigar Lake Uranium Deposit in Saskatchewan

Peter Vilks, J.J. Cramer, D.B. Bachinski,
D.C. Doern, and H.G. Miller

The Cigar Lake uranium deposit is located in northern Saskatchewan in the eastern part of the Athabasca Sandstone Basin. The deposit consists of a high-grade ore body (up to 65% U_3O_8) located at a depth of about 430 m. The ore occurs as an east-west-trending lens, 2 km long, 25 to 100 m wide, and 1 to 20 m thick. It is situated at the unconformity contact between the basal member of the Athabasca Sandstone formation and the underlying high-grade metamorphic rocks of the basement. The ore is surrounded by a clay-rich zone above and below. Above the ore body, there is a zone of altered sandstone, capped with a quartz-rich dome. The present-day groundwater flow is from the south to north through the lower sandstone unit directly above the unconformity.

Colloids (1 to 450 nm) and suspended particles (>450 nm) are being investigated to evaluate their effect on element transport through the uranium deposit (Vilks et al. 1988). As a prerequisite to understanding particle transport, research has concentrated on characterizing the size distribution, concentration, composition, and natural radionuclide content of particles throughout representative parts of the uranium deposit. Water samples (50 L) were filtered under a nitrogen atmosphere by tangential-flow ultrafiltration, which separated samples into four size fractions, consisting of dissolved species, colloids between 1 and 10 nm, colloids between 10 and 450 nm, and suspended particles larger than 450 nm. Particle concentrates were filtered through a series of Nuclepore membranes to determine the concentration and size distribution of particles larger than 10 nm. Scanning electron microscopy with energy dispersive X-ray analysis (SEM/EDX) was used to study the composition, size and morphology of

EFFECT OF BOREHOLE FLUSHING

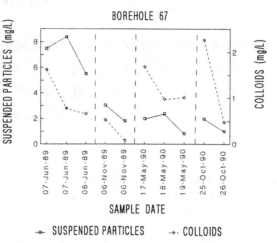

Figure 1.   The variation of colloid and suspended particle concentrations as a function of continued sampling from borehole 67, which collects water from the altered sandstone above the ore deposit. The dashed vertical lines are used to separate sampling periods.

particles larger than 100 nm. X-ray diffraction (XRD) was used to identify particle mineralogy. Information on particle compositions was obtained from chemical and radiochemical analysis of filtered water and particle concentrates.

Piezometers located in various parts of the deposit have been periodically sampled for colloids since 1986, using a squeeze pump with flow rates between 2 and 25 L/h. In a given sampling period, up to three colloid samples were taken from some boreholes, with the total volume of pumped water amounting to as much as 1000 L. In most boreholes the concentrations of suspended particles decreased during the collection of the first 350 L. Although these concentrations varied during subsequent sampling periods, they did not recover to the original values. On the other hand, colloid (10 to 450 nm) concentrations could not be permanently reduced by continued borehole flushing. For example, during a given sampling period, the colloid concentrations decrease with progressive sampling (Figure 1). However, colloid concentrations did recover during the time (4 to 6 months) between sampling periods. These results highlight the need for multiple sampling to obtain meaningful particle concentration data from the subsurface.

Particle size distributions, expressed as milligrams solid per liter of groundwater, showed that Cigar Lake groundwaters contain particles in all sizes ranging from 10 nm to slightly larger than 10 μm. Average colloid concentrations in groundwaters throughout the deposit ranged from 0.1 to 8 mg/L and did not show a distinct correlation with geological formation. Only water collected from borehole 199, intersecting a major fracture zone in the basement, contained unusually high colloid concentrations. Average concentrations of suspended particles varied from 1 to 144 mg/L and were determined more by the integrity of the local rock than by geologic formation. For example, parts of the ore zone consist of friable clays, which are easily mobilized by groundwater flow.

The SEM/EDX and XRD analyses have shown suspended particles to consist of typical fracture lining minerals, such as quartz, micas, clays, chlorite and amorphous Fe-Si oxyhydroxide. Some organic material in the form of bacteria and unidentified masses has been observed. The bulk compositions of colloid samples in the 1 to 10 nm and the 10 to 450 nm size ranges were determined by chemical analyses of particle concentrates. This method does not distinguish individual types of colloids, and it is not very sensitive to some elements, such as Si, which may have a high dissolved concentration. Colloids (1 to 450 nm) consist of organics, carbonates, iron oxides, and alumino silicates. The iron oxides are found mainly with the larger colloids (10 to 450 nm) and the suspended particles, suggesting that small iron colloids are not stable. The broad range of colloid compositions from the ore zone indicates that it may be impossible to identify the source of colloids within the deposit based on the major element composition of colloid samples.

One of the objectives of this study is to find evidence of colloid migration away from the ore body through geologic time. This can be achieved by finding a trace element or naturally occurring radioisotope that could be used to identify colloids originating from the ore body. Uranium is a potential candidate since in parts of the ore zone its concentrations associated with colloids and particles are higher than in the clay and sandstone surrounding the ore. Expressed in terms of micrograms per gram of particle, the colloids and suspended particles in the ore zone may contain up to 2000 μg/g U, while in the rest of the deposit U concentrations on particles are usually well below 600 μg/g.

The usefulness of U or any other element as a geochemical marker also depends upon whether the element can detach or desorb from particles during transit. The data for dissolved and particulate uranium can be used to calculate field-derived distribution coefficients (R_d) between particles and groundwater. The wide range of field-derived R_d observed for U in a given borehole indicate that U in particulate form may not be in equilibrium with groundwater. For example, in the clay/ore contact, the R_d values varied from 2000 to 800,000 mL/g. Therefore, U and other radionuclides, such as Th, may be potential markers for particle transport.

To calculate radiocolloid formation by radionuclide sorption onto colloids one must quantify the sorption process. The problem with using R_d values obtained from natural element distributions in the field is that elements may be attached to particles by mechanisms other than sorption. At the time of groundwater collection, a sorption experiment was carried out at Cigar Lake with natural colloids to validate the existing data on field derived R_d. Fifty-liter samples of water from holes 79, 197, and 220 were spiked with Sr, Cs, Zn, U, Ni, and Co. After about 2 h, these samples were filtered in the usual manner to determine element distributions between colloids (10 to 450 nm), suspended particles and groundwater. The distribution coefficients obtained from these sorption experiments were usually within the lower range of values measured for field-derived R_d, obtained from natural element distributions. Therefore, the existing data for field-derived R_d values could be used for predicting radiocolloid formation, provided that only the lower range of observed values are used.

This research is part of the Canadian Nuclear Fuel Waste Management research program supported by Ontario Hydro and the Candu Users Group. The studies at Cigar Lake, along with those at a number of uranium deposits throughout the world, are focused on understanding processes that have occurred in nature and that could occur in various nuclear fuel waste disposal concepts (Cramer 1986; Cramer et al. 1987).

REFERENCES

Cramer, J.J., A natural analog for a fuel disposal vault, In Proceedings of the 2d International Conference of Radioactive Waste Management, Canadian Nuclear Society, September 7-11, 1986, pp 697-702.

Cramer, J.J. and Vilks, P., JPA: Larocque: Near-field analog features from the Cigar Lake uranium deposit, in Come, B. and Chapman, N.A., (eds), CEC Symposium on Natural Analogues in Radioactive Waste Disposal, Brussels, April 28-30, 1987, EUR-11037, Comm. Europ. Com. Publication, 1987, pp 59-72.

Vilks, P., Cramer, J.J., Shewchuk, T.A., JPA: Larocque: Colloid and particulate matter studies in the Cigar Lake natural-analog program, *Radiochimica Acta* 44/45, 305-10, 1988.

CHAPTER 55

Visualization of Colloid Transport During Single and Two-Fluid Flow in Porous Media

Jiamin Wan and John L. Wilson

Although there are extensive studies of particle transport in porous media, few experimental studies have focused on pore scale behavior, especially with visual observations. Nor have many studies examined the role of gas-aqueous interfaces. In this research we explored these issues: (1) the behavior of particle attachment/detachment onto/from the solid surface at the pore scale and; (2) the role of the gas-aqueous interface on particle transport and particle mobilization.

In subsurface environments, gas as a major phase exists in the vadose zone above the water table. Gas bubbles can also be found in the saturated zone below. The role of gas-water interfaces on transport phenomena in porous media has received little attention. However, the interfacial properties have been intensively studied in the field of flotation. Flotation mechanisms (Derjaguin et al. 1981; Lekki and Laskowsky 1976; Hornaby and Leja 1982) consist of two parts: the attachment of finely ground minerals to the surfaces of gas bubbles and the migration of the fine-laden bubbles in aqueous phase. The process is governed by the surface properties of fines and by promoters which induce and increase adherence between fines and air bubbles. The mechanism of attachment of fines on air bubbles was suggested as the hydrophobicity. Goldenberg et al. (1989) also noted a strong adhesion of clay minerals to air bubbles in solutions with very high ionic strength. The mechanism they suggested is the compression of a diffuse electrical double layer.

Our experiments were conducted in glass micromodels, which were created by etching a pore network pattern onto two glass plates which were then fused together in a furnace. Micromodels offer the ability to visually observe fluid and particle behavior on a bulk scale and within individual pores. Although the pore network is only two-dimensional, the pores have a complex three-dimensional structure (Wilson et al. 1990; Wan and Wilson 1991). Similar to the structure of a natural porous medium, a typical network consists of pore bodies, pore throats, and pore wedges. The wedges represent the corners of pore bodies and throats. By analogy a circular pore throat has no wedge, whereas a triangular-shaped throat would have three wedges in its three vertices. Three different micromodels were used. One had a homogeneous network with a low respect ratio (pore throats that were not much smaller than pore bodies). The second was a homogeneous network with a high respect ratio (pore throats were much smaller than pore bodies). Another micromodel was a network with preferential flow channels representing macropores or fractures. Pore sizes varied from 2 to 600 μm, and the total pore volume for each micromodel varied from 0.07 to 0.25 mL. The models had dead zones. These were portions of the pore space which are not well connected to the flow field, or they were stagnant zones in which velocities were small because of pore scale fluid mechanics.

Two kinds of polystyrene latex microspheres were used. One was a negatively charged carboxylated particle and the other was a positively charged particle with amino function groups. Their sizes were 0.92 and 1.02 μm, respectively. TiO_2 particles (about 0.3 μm) were used in solutions of 3.9 and 10.3 pH, in which the particles behaved positively and negatively, respectively. The experiments were conducted with an aqueous solution of 0.001 M NaCl and air. During flow the volumetric displacement was set at 0.15 mL per hour, and fluids were injected with a syringe pump. The micromodel was set horizontally under a microscope. The experiments were observed through the microscope and recorded by a video camera.

The experimental procedure was:

1. Saturate a clean micromodel with the NaCl solution, trapping some air bubbles as a residual phase.
2. Inject a dilute particle suspension at a constant rate for 20 or 30 pore volumes. Observe attachment of the particles to the glass pore surface.
3. Inject a particle-free aqueous solution to replace the suspension at the same rate (0.15 mL/h) for about five pore volumes. Observe selected detachment of particles and evolution of a steady-state population of attached particles.
4. Replace the aqueous phase with air, reducing water to a residual saturation. Observe detachment of particles from pore surface onto the air–water interface during the interface migration.
5. Inject the aqueous phase to remove the air. Observe the particles adsorbed on the interfaces moving out from the model.

Observations revealed different degrees of particle attachment and detachment on/from the solid surface, where more positively charged particles were adsorbed on pore surface than negatively charged ones, also TiO^2 particles. Attached negative particles were easier to detach than positive particles under the same experimental conditions, especially the negatively charged TiO_2 particles. The observations were conceptually consistent with DLVO theory. The particle spatial distribution patterns during attachment and detachment were similar for oppositely charged particles. In pore bodies, adsorbed particles were uniformly dispersed; the concentration was higher on the bottom than on the ceiling of the horizontal micromodels due to sedimentation. In pore throats, preferential attachment was observed. In large throats, whose sizes were not much smaller than that of the pore bodies, the concentration of particles was greater than in the pore bodies. In the small throats, whose sizes were much smaller than the body sizes, particles accumulated and finally blocked the throats. In very fine throats straining was observed. The flow velocity is higher in pore throats than in pore bodies, and there is greater propensity for interception. Particles have greater momentum and it becomes easier for negatively charged particles to penetrate the energy barrier. The pattern of attachment for negatively charged TiO_2 even mimicked the flowlines. In pore wedges significant preferential attachment was observed. Flow velocity is lower in wedges and, it gives rise to greater probability of sedimentation and Brownian diffusion. In dead zones a large concentration of particles was observed. Sedimentation and Brownian diffusion dominate. The details of these phenomena depend upon particle properties, pore structure, surface roughness, solution chemistry, and flow conditions.

It was observed that the air–water interface efficiently adsorbed particles. Both positively and negatively charged polystyrene and TiO_2 particles were preferentially adsorbed onto the interface relative to the glass surface. This phenomenon is not well understood (Wan and Wilson 1991). Toxvaerd (1973) suggests that the presence of a gas–water interface gives rise to an increase in the degree of structuring of water. In the boundary layer the water is denser than in the rest of the system. In this region, there is a mean orientation of the water molecules with the positive hydrogen toward the bulk solution and the negative oxygen towards the air. The dipolar contribution of this average orientation of water molecules in the surface zone results in a positive surface potential. But in our experiments both positively and negatively charged particles adsorbed onto the gas–
water interface, and the positively charged one had a stronger affinity. Particles appeared to randomly distribute on the air–water interface, although they sometimes were more concentrated on the downstream side of bubbles, and sometimes appeared to concentrate in areas of greater curvature.

Once adsorbed onto the gas–water interface the adsorbed particles were not easily dislodged back into the aqueous phase. Increasing the flow velocity caused some particles on the interface to move faster either along the interface, or oscillating in place. Other particles were stationary. Almost no particles could be removed from the interface by flow unless, for example, an entire air bubble was moved.

During the process of injecting air into the pore network the air-water interface very powerfully stripped the adsorbed particles from solid surface. The interface is energetically preferred by the particles. There was also local velocity jumps as each pore drained and the interface entered (Haines jumps) that may have played a role. The detachment of particles by the air-water interface was most effective in pore bodies and large throats. It was not very effective in deep pore wedges, fine throats, and dead zones since air was not able to reach there. The detached particles were then carried away by the interface during the subsequent injection of water.

In conclusion it appears that adsorption of colloidal particles was greater in pore wedges than in pore throats, and in pore throats greater than in pore bodies. Continuous sorption resulted in blocking of small throats. Desorption was in an opposite order. Particles attached in the pore bodies were much easier to detach by flow than those attached anywhere else. Particles blocking the small throats were not removed by flow unless the flow direction was reversed. The gas–water interface efficiently adsorbed all of the different kinds of particles used in the experiments. Once particles were attached onto the air–water interface surface, they can not be removed unless the air bubble was removed. Increasing flow velocity caused some particles to oscillate rapidly on the gas–water interface. During injection of air the gas–water interface detached most of the particles that the interface reached. The interface carried these particles away from the system during injection of water. Multiple air-water replacement cycles removed most of the attached particles from the pore space.

REFERENCES

Derjaguin, B.V. and Dukhin, S.S., Kinetic theory of the flotation of fine particles, in Developments in Mineral Processing (Proc. XIII Int. Miner. Processing Congr., Warsaw, Poland), eds. Laskowski, J., Elsevier: Amsterdam; Polish Scientific Publishers, Warsaw 2,21, 1981.

Hornsby, D. and Leja, J., Selective flotation and its surface chemical characteristics, *Surface Colloid Sci* 12:217-314, 1982.

Goldenberg, L.C., Hutcheon, I., and Wardlaw, N., Experiments on transport of hydrophobic particles and gas bubbles in porous media, *Transport in Porous media* 4:129-145, 1989.

Lekki, J. and Laskowsky, J., Dynamic interaction in particle-bubble attachment in flotation, in Kerker, M., (ed.), *Colloid and Interface Science*, Vol. 4, Academic Press, New York, 1976.

Sutcliffe, W.H., Baylar, E.R., and Menzel, W.D., Sea surface chemistry and Langmuir circulation, *Deep-sea Res* 10:233-243, 1963.

Toxvaerd, S., Surface structure of simple fluids, *Prog Surface Sci* 9:189-220, 1973.

Wan, J.M. and Wilson, J.L., Colloid transport and the gas-water interface in porous media, in *Colloid and interfacial aspects of groundwater and soil cleanup*, ACS Books Department, 1991.

Wilson, J.L., Conrad, S.H., Mason, W.R., Peplinski, W., and Hagan, E., Laboratory investigation of residual liquid organic, Robert SK, Environmental Research Laboratory; EPA/600/6-90/004, 1990.

Appendix A

Agenda of Meeting

MANTEO III

CONCEPTS IN MANIPULATION OF GROUNDWATER COLLOIDS
FOR ENVIRONMENTAL RESTORATION

October 15-18, 1990
North Carolina Aquarium
Manteo, North Carolina

AGENDA

OBJECTIVE OF THE MEETING: *To review current understanding of colloid behavior in subsurface systems and to examine the feasibility of conceptual strategies for manipulating colloids for remediation or containment of hazardous wastes.*

MONDAY EVENING, October 15, 1990

7:00 - **Registration, Reception and Mixer** (Elizabethan Inn)

TUESDAY, October 16, 1990

8:00 - **Registration**
8:15 - **Welcoming Remarks**
8:20 - **DOE Subsurface Science Program**
       Frank J. Wobber (US Department of Energy)
8:25 - **Meeting Objectives and Organization**
       John F. McCarthy (Oak Ridge National Laboratory)

SESSION I: MOBILIZING AND DEPOSITING COLLOIDS
   Objective: Can chemical modification of subsurface systems be used to reliably manipulate colloid stability in situ?

8:45 - **Moderator's Overview**
       John C. Westall (Oregon State University)
9:15 - **Particle Deposition in Porous Media in the Presence of Repulsive Double Layer Force**
       Chi Tien (Syracuse University)

9:40 - **Aggregation Kinetics and Fractal Structure of Colloidal Oxides**
T. David Waite and R. Amal (Australian Nuclear Science and Technology Organization)

10:05 - **BREAK**

10:30 - **Field Manipulations of Organic and Inorganic Colloids in a Sandy Aquifer**
John F. McCarthy, Liyuan Liang, Philip M. Jardine (Oak Ridge National Laboratory) and Thomas M. Williams (Clemson University)

10:55 - **Hydrogeochemical Factors Influencing Colloids in Waste Disposal Environments**
Philip M. Gschwend (Massachusetts Institute of Technology)

11:20 - **DISCUSSION**
Moderator: Philip M. Gschwend (Massachusetts Institute of Technology) and John M. Zachara (Pacific Northwest Laboratory)

12:15 - **LUNCH**

SESSION II: BIOCOLLOID MOBILITY

Objective: What controls mobility of microorganisms in porous and fractured media, and can this mobility be manipulated to improve delivery of microbes for bioremediation?

1:30 - **Moderator's Overview**
Ronald W. Harvey (US Geological Survey)

2:00 - **Movement of Microorganisms Through Unconsolidated Porous Media Without Fluid Flow**
Michael McInerney (University of Oklahoma)

2:20 - **Cells and Surfaces in Groundwater Systems**
Robert E. Martin (Lyonnaise des Eaux, France), Edward J. Bouwer, and Linda Hanna (Johns Hopkins University)

2:40 - **Deposition Behavior of Bacteria in Column Systems**
Huub Rijnaarts and Alexander Zehnder (Agricultural University, The Netherlands)

3:00 - **BREAK**

3:30 - **Physical and Chemical Controls on Advective Transport of Bacteria**
Aaron Mills, D. E. Fontes, G. M. Hornberger, and J. S. Herman (University of Virginia)

3:50 -   **Modelling Biocolloid Transport in Porous Media:**
        Chemical Aspects and Reversibility
        Roger Bales (University of Arizona)

4:10 -   **DISCUSSION**
        Moderator: Edward J. Bouwer (Johns Hopkins University)

5:00 -   **POSTER SESSION AND SOCIAL HOUR(S)**

WEDNESDAY, October 17, 1990

SESSION III: MANIPULATING COLLOIDS TO CHANGE AQUIFER PERMEABILITY

Objective: Will changes in colloid stability affect the physical structure of a formation to either increase permeability (to enhance bioremediation) or clog pores to create in situ barriers?

8:30 -   **Moderator's Overview**
        James R. Hunt (University of California, Berkeley)

9:00 -   **Influence of Biofilm Accumulation on Porous Media Hydrodynamics**
        Al B. Cunningham (Montana State University)

9:20 -   **Biomass Manipulation to Control Pore Clogging in Aquifers**
        Peter R. Jaffe (Princeton University)

9:40 -   **Colloid Entrainment, Transport, and Entrapment in Two Phase Flow in Porous Media**
        M. M. Sharma (University of Texas)

10:00 - **BREAK**

10:30 - **Effects of pH on Fines Migration and Permeability Reduction**
        H. Scott Fogler and Ravi Vaidya (University of Michigan)

10:50 - **Effects of Cellular Polysaccharide Production on Cell Transport and Retention in Porous Media**
        Ray Lappan, and H. Scott Fogler (University of Michigan)

11:10 - **DISCUSSION**
        Moderator: H. Scott Fogler (University of Michigan)

12:00 - **LUNCH**

SESSION IV: MODIFIED COLLOIDS, SURFACTANTS, AND EMULSIONS

Objective: Surfactants can modify surface binding sites on colloids, or can promote formation of colloidal emulsions;

can these be used to control colloid mobility or to increase recovery of sorbed contaminants by pump and treat methods?

1:00 - **Moderator's Overview**
Richard G. Luthy (Carnegie Mellon University)
1:30 - **Surfactant Mobilization of NAPL Residuals in Subsurface Systems**
Kim F. Hayes (University of Michigan)
1:50 - **Adsorption of Ionic Surfactants and Their Role in Colloid Stability**
Douglas W. Fuerstenau (University of California, Berkeley)
2:10 - **Reactions of Anionic Surfactants in Saturated Soils and Sediments: Precipitation and Micelle Formation**
Chad T. Jafvert (USEPA, Athens)
2:30 - **Clay Modification for Environmental Restoration and Waste Management**
Stephen A. Boyd (Michigan State University)
2:50 - **BREAK**
3:20 - **DISCUSSION**
Moderator: John C. Westall (Oregon State University)

SESSION V: COUPLED PROCESSES IN COLLOID MANIPULATIONS

Objective: Consolidate previous discussions on how proposed manipulation strategies will be influenced by the multiply-interacting microbiological, chemical, and hydrologic factors in natural systems; is the cure worse than the disease?

4:00 - **Moderator's Overview**
John M. Zachara (Pacific Northwest Laboratory)
4:20 - **DISCUSSION**
Moderator: John M. Zachara
5:00 - **POSTER SESSION AND SOCIAL HOUR(S)**

THURSDAY, October 18, 1990

Session VI: Working Group Consensus on Manipulation Concepts and Strategies

Objective: Participants will divide into working groups to summarize discussions on each topic, identify strengths and limitations of different conceptual strategies, and identify long-term research needs for achieving the most feasible strategies.

8:00 -	Charge to the Working Groups
8:15 -	Working Group Meetings (Sessions I and II)
9:45 -	**BREAK**
10:15 -	Working Group Meetings (Sessions III and IV)
11:30 -	Reassembled Group Discussion and Conclusions
12:30 -	**ADJOURNMENT**

For additional information, please contact the organizer:
John F. McCarthy
Environmental Sciences Division
Oak Ridge National Laboratory
P.O. Box 2008
Oak Ridge, Tennessee 37831-6036, USA
Phone: (615) 576-6606 / Fax: (615) 576-8543

Appendix B

List of Participants

Dr. Gary Amy
Civil, Environmental, and
  Architectural Engineering
University of Colorado
Boulder, CO 80309
(303) 492-6274
(FAX) 303 492-7315

Dr. Paul R. Anderson
Department Environmental
  Engineering
Illinois Institute of Technology
3201 South State Street
Chicago, IL 60616
(312) 567-3531

Dr. Charles B. Andrews
SSP & A, Suite 290
12250 Rockville Pike
Rockville, MD 20852
(301) 468-5760
(FAX) 301 881-0832

Dr. Roger Bales
Department Hydrology & Water
  Resources
University of Arizona
Tucson, AZ 85721
(602) 621-7113
(FAX) 602-621-1422

Dr. Mark Benjamin
Department of Civil Engineering
FX-10
University of Washington
Seattle, WA 98195
(206) 543-7645
(FAX) 206-543-1543

Dr. Ed Bouwer
Department of Geography and
  Environmental Engineering
The Johns Hopkins University
34th and Charles, Ames Hall
Baltimore, MD 21218
(301) 338-7437
(FAX) 301 338-8996

Dr. Steven A. Boyd
Michigan State University
Department of Crop and Soil
  Science
East Lansing, MI 48824-1325
(517) 353-3993
(FAX) 517 353-5174

Dr. Marilyn Buchholtz ten Brink
L202 Earth Science Department
Lawrence Livermore National
  Laboratory
P.O. Box 808
Livermore, CA 94550
(415) 423-7662
(FAX) 415 423-1997

Dr. Steve Cabaniss
Department of Chemistry
Kent State University
Kent, OH 44242
(216) 672-3731
(FAX) 216-672-3816

Dr. Tom Cronk
Oak Ridge National Laboratory
P.O. Box 2567
Grand Junction, CO 81502
(303) 248-6265
(FAX) 303-248-6147

351

Dr. Al B. Cunningham
Center for Interfacial Microbial
  Process Engineering
Montana State University
409 Cobleigh Hall
Bozeman, MT 59717
(406) 994-4770
(FAX) 406 994-6098

Dr. Mervyn Frank Daniel
Shell Development Company
Westhollow Research Center
P.O. Box 1380
Houston, TX 77251-1380
(713) 493-8427/8303
(FAX) 713 493-7705

Dr. Menachem Elimelech
4532-D Boelter Hall
Civil Engineering Department
University of California
Los Angeles, CA 90024-1593
(213) 825-1774
(FAX) 213 206-2222

Dr. Edward H. Essington
P.O. Box 1663, MS-J495
Los Alamos, NM 87545
(505) 667-3057
(FAX) 505 665-3866

Dr. Kevin J. Farley
Environmental Systems
  Engineering
Clemson University
Clemson, SC 29634-0919
(803) 656-4202
(FAX) 803 656-0672

Dr. H. Scott Fogler
Department of Chemical
  Engineering
374 Dow Building
University of Michigan
Ann Arbor, MI 48109
(313) 769-0362

Dr. David Fontes
Department of Environmental
  Sciences
Clark Hall
University of Virginia
Charlottesville, VA 22903
(804) 924-7761
(FAX) 804 982-2137

Dr. D.W. Fuerstenau
Department of Materials Science
  and Mineral Engineering
University of California
Berkeley, CA 94720
(415) 642-3826
(FAX) 415 642-6623

Dr. Jeffrey S. Gaffney
Argonne National Laboratory
Environmental Research Division
Building 203
Argonne, IL 60439
(708) 972-5178
(FAX) 708 972-5498

Dr. M.P. Gardiner
Room 102 Bio 30 Experimental
Studies Department
Harwell Laboratory
Oxfordshire OX11 ORA
UNITED KINGDOM
44-0235-821111 x4981
(FAX) 44-0235-43492

Dr. Scott Grace
Department of Energy-Rocky
 Flats Office
P.O. Box 928 - ERD
Golden, CO 80402-0928
(303) 966-7199
(FAX) 303 966-2256

Dr. Philip M. Gschwend
Department of Civil Engineering,
 48-415
Massachusetts Institute of
 Technology
Cambridge, MA 02139
(617) 253-1638
(FAX) 617 258-8850

Dr. Ronald W. Harvey
U.S. Geological Survey
325 Broadway
Boulder, CO 80303-3328
(303) 541-3034
(FAX) 303-447-2505

Dr. Kim F. Hayes
University of Michigan
Department of Civil Engineering
Ann Arbor, MI 48109-2125
(313) 763-9661
(FAX) 313 764-4292

Dr. Remy J-C Hennet
SSP & A, Suite 290
12250 Rockville Pike
Rockville, MD 20852
(301) 468-5760
(FAX) 301 881-0832

Dr. Janet S. Herman
Department of Environmental
 Sciences
Clark Hall
University of Virginia
Charlottesville, VA 22903
(804) 924-7761
(FAX) 804 982-2137

Dr. James R. Hunt
Department Civil Engineering
University of California
Berkeley, CA 94704
(415) 642-0948
(FAX) 415 643-5264

Dr. Peter R. Jaffe
Department of Civil Engineering
 and Operations Research
Princeton University
Princeton, NJ 08544
(609) 258-4653
(FAX) 609 258-1270

Dr. Chad T. Jafvert
U.S. Environmental Protection
 Agency
Environmental Research Lab
College Station Road
Athens, GA 30613

Ms. Lou Jolley
Baruch Forest Sciences Institute
Clemson University
Box 596, Highway 17
Georgetown, SC 29442
(803) 546-1013
(FAX) 803 546-6296

Dr. Thomas Kulp
Lawrence Livermore Laboratory
P.O. Box 5507 L-524
Livermore, CA 94513
(415) 423-7839
(FAX) 422-8020

Mr. Ray Lappen
Department of Chemical
  Engineering
374 Dow Building
University of Michigan
Ann Arbor, MI 48109
(313) 764-4313

Dr. Liyuan Liang
Environmental Sciences Division
Oak Ridge National Laboratory
P.O. Box 2008
Oak Ridge, Tennessee 37831-6038
(615) 574-9387
(FAX) 615 576-8646

Dr. Leonard W. Lion
School of Civil & Environmental
  Engineering
Hollister Hall
Cornell University
Ithaca, NY 14853
(607) 255-7571
(FAX) 607 255-9004

Dr. Geoffrey Longworth
Isotope Geoscience Station
Building 7
AEA Industrial Technology
Harwell Laboratory
Oxfordshire OX11 ORA
UNITED KINGDOM
61-0235-432745
(FAX) 61-0235-432413

Dr. Richard G. Luthy
Department of Civil Engineering
Carnegie Mellon University
Pittsburgh, PA 15213-3890
(412) 268-2941
(FAX) 412 268-7813

Dr. Nancy A. Marley
Argonne National Laboratory
Environmental Research Division
Building 203
Argonne, IL 60439
(708) 972-5178
(FAX) 708 972-5014

Dr. Robert E. Martin
Lyonnaise des Eaux-Dumez
72 Avenue de la Liberte
9200 Nanterre
FRANCE
33-14695-5237
(FAX) 33-14695-5265

Dr. John F. McCarthy
Environmental Sciences Division
Oak Ridge National Laboratory
P.O. Box 2008
Oak Ridge, Tennessee 37831-6036
(615) 576-6606
(FAX) 615 576-8543

Dr. Michael J. McInerney
Department of Botany &
  Microbiology
University of Oklahoma
770 Van Vleet Oval
Norman, OK 73019-0245
(405) 325-6050
(FAX) 405 325-7619

Dr. Clarence A. Miller
Department of Chemical
  Engineering
Rice University
P.O. Box 1892
Houston, TX 77251
(713) 527-4904
(FAX) 713 524-5237

Dr. Aaron L. Mills
Department of Environmental
  Sciences
Clark Hall
University of Virginia
Charlottesville, VA 22903
(804) 924-7761
(FAX) 804 982-2137

Dr. Valerie Moulin
Centre d'Etudes Nucleaires,
  CEN-FAR
DCC/DSD/SCS/SGC
BP 6 92265 Fontenay-aux-Roses
Cedex
FRANCE
33-1465-47775
(FAX) 33-1425-31469

Dr. Larry R. Myer
Earth Sciences Division
Lawrence Berkeley Laboratory
Berkeley, CA 94720
(415) 486-6456
(FAX) 415 486-5686

Dr. James A. Novitsky
Dalhousie University
Department of Biology
Halifax, Nova Scotia
CANADA B3H 4J1
(902) 424-3515
(FAX) 902 424-3736

Dr. H. Eric Nuttall
University of New Mexico
Chemical and Nuclear Engineering
Farris Engineering Center
Room 209
Albuquerque, NM 87131
(505) 277-6112
(FAX) 505 277-0813

Dr. Terese M. Olson
Department of Civil Engineering,
  101 ICEF
University of California, Irvine
Irvine, CA 92717
(714) 856-7188
(FAX) 714 725-2117

Dr. Charles R. O'Melia
Department of Geography and
  Environmental Engineering
The John's Hopkins University
Baltimore, MD 21218
(301) 338-7102
(FAX) 301 338-8996

Dr. Dale L. Perry
Mail Stop 70A-1150
Lawrence Berkeley Laboratory
University of California
Berkeley, CA 94720
(415) 486-4819
(FAX) 415 451-4819

Dr. Robert W. Puls
R.S. Kerr Environmental
  Research Laboratory
U.S. Environmental Protection
  Agency
P.O. Box 1198
Ada, OK 74820
(405) 332-8800
(FAX) 405 332-8800

Dr. Shirley A. Rawson
Idaho National Engineering
   Laboratory
EG&G Idaho, Inc.
P.O. Box 1625, MS-2107
Idaho Falls, ID 83415-2107
(208) 526-8628
(FAX) 208-526-9822

Dr. Huub Rijnaarts
Department of Microbiology
Agricultural University
Hesselink van Suchtelenweg 4
6703 CT Wageningen
THE NETHERLANDS
31-8370-82105
(FAX) 31-8370-83829

Dr. Bill Rixey
Shell Development
P.O. Box 1390
Houston, TX 77251-1380
(713) 493-8313
(FAX) 713-493-8727

Dr. Wayne P. Robarge
North Carolina State University
Box 7619 Soil Science
Raleigh, NC 27695-7619
(919) 737-2600
(FAX) 919 737-7422

Mr. James E. Saiers
Department of Environmental
   Sciences
Clark Hall
University of Virginia
Charlottesville, VA 22903
(804) 924-7761
(FAX) 804 982-2137

Dr. Marcus A. Sen
British Geological Survey
Keyworth Nottingham, NG12 5GG
UNITED KINGDOM
44-06077-61111
(FAX) 44-06077-4841

Dr. M.M. Sharma
Department of Petroleum
   Engineering
University of Texas at Austin
Austin, TX 78712

Dr. Barry Smith
Fluid Processes Unit
British Geological Survey
Keyworth
Nottingham NG125GG
UNITED KINGDOM
44-06077-6111, ext. 3522
(FAX) 44 01144 06077 4841

Dr. D. Kip Solomon
Oak Ridge National Laboratory
Building 1504, MS-6351
Oak Ridge, TN 37831-6351
(615) 576-2950
(FAX) 615 574-4946

Dr. T.B. Stauffer
HQ AFESC/RDVC
Tyndall AFB, FL 32403-6001
(904) 283-4297
(FAX) 904 283-6499

Professor Chi Tien
411 Link Hall
Department of Chemical
  Engineering
Syracuse University
Syracuse, NY 13244
(315) 443-4050
(FAX) 315 443-4936

Dr. John E. Tobiason
Department of Civil Engineering
University of Massachusetts
Amherst, MA 01003
(413) 545-0685
(FAX) 413 545-0724

Dr. Ines R. Triay
Los Alamos National Laboratory
MS J514
Los Alamos, NM 87545
(505) 665-1755
(FAX) 505 665-4955

Dr. Brian E. Viani
Lawrence Livermore National
  Laboratory
P.O. Box 808
Livermore, CA 94550
(415) 423-2001
(FAX) 415 422-1002

Dr. Peter Vilks
Whiteshell Laboratories
Pinawa Manitoba, ROE 1LO
CANADA
(204) 753-2311
(FAX) 204 753-2455

Dr. T. David Waite
Australian Nuclear Science &
  Technology
Organization, Lucas Heights
  Research Laboratories
New Illawarra Road
Lucas Heights, NSW, Private Mail
Bag 1
MENAI NSW 2234, AUSTRALIA
61 (02) 543-3111
(FAX) 61 (02) 543-7536

Ms. Jiamin Wan
Hydrology Program
Geoscience/Hydrology
Campus Station
Socorro, NM 87801
(505) 835-5465
(FAX) 505 835-6329

Dr. Peter Warwick
Department of Chemistry
  Loughborough
University of Technology
  Loughborough
Leicestershire
UNITED KINGDOM
44-0509-222588
(FAX) 44-0509-233163

Dr. John C. Westall
Chemistry Department
Oregon State University
Corvallis, OR 97330
(503) 737-2591
(FAX) 503 737-2062

Dr. Thomas M. Williams
Baruch Forest Science Institute
Box 596
Georgetown, SC 29442
(803) 546-6318
(FAX) 803 546-6296

Dr. John L. Wilson
Hydrology Program
Geoscience/Hydrology
Campus Station
Socorro, NM 87801
(505) 835-5308
(FAX) 505 835-6329

Dr. Caroline S. Wittwer
Department of Geology and
  Geophysics
University of California, Berkeley
Berkeley, CA 94720
(415) 642-3584
(FAX) 643-9980

Dr. Frank J. Wobber
Program Manager
Subsurface Science Program
Office of Energy Research (ER-74)
U.S. Department of Energy
Washington, DC 20545
(301) 353-5549
(FAX) 301 353-5051

Dr. John M. Zachara
Earth Sciences Department
Battelle Pacific Northwest
  Laboratory
P.O. Box 999
Richland, WA 99320
(509) 375-2993
(FAX) 509 375-2718

AUTHOR INDEX

Philip M. Jardine, Environmental Sciences Division, Oak
   Ridge National Laboratory, Oak Ridge, Tennessee *35, 309*

Lou Jolley, Baruch Forest Sciences Institute, Clemson
   University, Georgetown, South Carolina *263*

P.M. Kearl, Oak Ridge National Laboratory, Grand Junction, Colorado . . *211*

Ruben M. Kretzschmar, North Carolina State University,
   Soil Science, Raleigh, North Carolina *253*

Ray Lappan, Department of Chemical Engineering, University
   of Michigan, Ann Arbor, Michigan *129*

Ann T. Lemley, College of Human Ecology, Cornell University,
   Ithaca, New York *275*

Liyuan Liang, Environmental Sciences Division, Oak Ridge
   National Laboratory, Oak Ridge, Tennessee *35, 263, 309*

Jong-Choo Lim, Department of Chemical Engineering, Rice
   University, Houston, Texas *269*

Leonard W. Lion, School of Civil & Environmental Engineering,
   Cornell University, Ithaca, New York *275, 325*

Gary M. Litton, Department of Civil Engineering, University of
   California-Irvine, Irvine, California *283*

Houmao Liu, Department of Hydrology and Water Resources,
   University of Arizona, Tucson, Arizona *191*

Zhongbao Liu, Department of Civil Engineering, Carnegie
   Mellon University, Pittsburgh, Pennsylvania *213*

Geoffrey Longworth, Isotope Geoscience Station, AEA
   Industrial Technology, Harwell Laboratory, Oxfordshire,
   United Kingdom *289*

Richard G. Luthy, Department of Civil Engineering, Carnegie
   Mellon University, Pittsburgh, Pennsylvania *153, 181, 213*

Johannes Lyklema, Department of Microbiology, Agricultural
   University, Wageningen, The Netherlands *73*

Brian R. Magee, Roy F. Weston, West Chester, Pennsylvania *275*